KNOW THE LANDSCAPE

Farms and Fields

Susanna Wade Martins

B.T. BATSFORD LTD · LONDON

First published 1995
© Susanna Wade Martins, 1995

All rights reserved. No part of this publication may be
reproduced in any form or by any means, without permission from the Publisher.

Typeset by Best-set Typesetter Ltd, Hong Kong
and printed in Great Britain by Butler and Tanner, Frome

Published by B.T. Batsford Ltd
4 Fitzhardinge Street, London W1H 0AH

A catalogue record for this book is
available from the British Library

ISBN 0 7134 6790 8

Contents

	List of illustrations	4
	Acknowledgements	6
	Introduction	7
	The evolving landscape	9
1	The beginnings of agriculture – the earliest fields	9
2	From the Romans to Domesday Book	22
3	Early medieval agriculture and its field systems	30
4	The period of medieval expansion	40
5	Contraction and consolidation – the late medieval landscape	53
6	1500–1700 The rise of the gentry farm	62
7	An agricultural revolution? The landscape evidence	75
8	The Welsh and Scottish experience	91
9	'High farming' – the heyday of the landed estate	101
10	Stagnation and diversification – the last 120 years	113
	The reclaimed landscape	121
11	The reclaimers	121
12	The Duke of Bedford in the fens	124
13	The transformation of the county of Sutherland	133
	Epilogue	146
	Notes to the text	150
	Bibliography	153
	Index	157

Illustrations

1. Bronze Age fields on Dartmoor, Devon — 10
2. Bronze Age ditches at Fengate, Cambridgeshire — 11
3. Wicca Farm, Cornwall — 12
4. Iron-Age huts at Butser Hill, Hampshire — 14
5. Strip lynchets at Askerswell, Dorset — 15
6. 'Celtic' fields, Fyfield Down, Wiltshire — 15
7. Roman walled enclosures at Wharfedale, Dorset — 17
8. Roman settlement at Crosby Ravensworth, Cumbria — 18
9. Roman fields around Thurrock, Essex — 19
10. The Fen Causeway, Norfolk and the rectangular fields of a Roman Farm — 20
11. Reconstruction of Anglo-Saxon village at West Stow, Suffolk and a plan of the fields around the settlement — 24
12. The village of Wharram Percy, Yorkshire showing the Iron Age and Romano-British sites — 26
13. Reconstruction of the open fields at Butterwick, East Yorkshire in 1563 — 27
14. Plan of Longham, Norfolk showing population expansion and shifts — 28
15. The village of Milburn, Cumbria — 31
16. Planned village at Appleton-le-Moors, Yorkshire — 31
17. Wimpole Park, Cambridgeshire, showing ridge and furrow — 33
18. The open fields of Laxton, Nottinghamshire showing open fields — 34
19. 1625 map of Laxton — 36
20. Laxton farms — 37
21. Map of Laughton Parish, Kent — 38
22. Reconstruction of Evesham Abbey Grange, Gloucestershire — 43
23. The barley barn at Cressing Temple, Essex — 45
24. Reconstruction of the Cistercian barns of St Leonards and Great Coxwell, Oxfordshire — 46
25. The medieval fenland parish of Marshland, Norfolk — 48
26. Map of part of Langdale, Cumbria — 49
27. The deserted village of Grenstein, Norfolk — 55
28. Reconstruction of 14th century Wharram Percy — 58
29. A Dartmoor longhouse at Oldsbriun, Devon — 59
30. A cruck barn, Wharram Percy — 60
31. Swaledale, Yorkshire — 66
32. Water Meadows, Nadder Valley, Wiltshire — 67
33. Old and New Bedford rivers, Cambridgeshire — 69
34. Channonz Hall, Norfolk — 70
35. Cressingham Manor, Norfolk — 70

List of illustrations

36	A farm on the Norfolk Broads, edge at Burlingham	71
37	Hollowbank, and bankbarn at Satherwaite, Cumbria	73
38	The Old House at Allendale, Northumberland	74
39	A map of Corpusty, Norfolk	76
40	Map showing re-laid out estate farms in Weasenham and Wellingham, Norfolk	77
41	An enclosure road at North Elmham, Norfolk	80
42	The pumping station at Ten Mile Bank in the Norfolk fens	81
43	Thorndon Hall, Essex	84
44	Farm buildings at South Creake, Norfolk	84
45	The Home Farm at Roundway, Wiltshire	85
46	The Home Farm at Holkham, Norfolk	86
47	The Home Farm at Wimpole, Cambridgeshire	87
48	Chollerton Farm, Northumberland	88
49	Maramette Farm, Yorkshire	89
50	Longhouse at Cilewent, Powys	93
51	Pants' map of Craigyloach, Perthshire	94
52	Model cattle steadings at Aden, Grampian	97
53	Grain-drying kiln at Sibster, Caithness	98
54	Wheel house for a horse gin at Sibster	98
55	Run rig fields on the Isle of Eigg, Inner Hebrides	99
56	Laterfield boundaries on the Isle of Eigg	100
57	Digging bush drains at Bedingham Farm, Suffolk	102
58	Drainage tile-making machines	104
59	Model farm buildings at Holkham, Norfolk	105
60	Newbridge Farm, Nordelph, Norfolk	106
61	Manor Farm, Crimplesham, Norfolk	108
62	Internal fittings at Crimplesham	109
63	Covered cattle yards at East Harptree, Avon	110
64	Cattle courts at Eastfield Farm, East Lothian	111
65	Cattle courts at Eastfield Farm	111
66	Circular sheep fold on the Cheviot Hills, Northumberland	114
67	Fanks at Edderton, Easter Ross	115
68	Flat Cambridgeshire countryside – changes over the last 50 years	119
69	Land-use changes in Huntingdonshire, 1937–60	120
70	Changes in Dorset, 1937–60	120
71	Location map of Thorney Level, Cambridgeshire	125
72	Thorney Level, pattern of farms and ditches	126
73	Thorney Level in 1749	127
74	Estate workshops at Thorney	129
75	Mr Goodman's Farm, Thorney	130
76	Knarr's Cross Farm, Thorney	131
77	A farmhouse on the Thorney Estate	132
78	Location map of the Sutherland Estate	134
79	Kintradwell, Sutherland, 1722	138
80	Kintradwell, 1880	139
81	Kintradwell farm buildings	139
82	Farm buildings at Craigkaig, Sutherland	141
83	The new town of Helmsdale, Sutherland, 1880s	142
84	Torrish sheep farm, Sutherland	143
85	Steadings at Clynelish, Sutherland	145

Acknowledgements

This book is the result of over twenty years' work in the field and owes much to a wide variety of people, both academics and farmers, all of whom have freely shared their expertise with me. Much of my recent research has only been possible through the support of the Centre of East Anglian Studies at the University of East Anglia where I have been a Research Associate for the past eight years, and colleagues there, especially Dr Richard Wilson, Dr Tom Williamson and Peter Murphy have all given much helpful advice. The Field Archaeology Division of the Norfolk Museum Service, particularly Derek Edwards of the air photography collection and Edwin Rose of the Sites and Monuments Record have also been very helpful. The staff of Norfolk Record Office have, as always, been very patient with my enquiries as have those at the Cambridgeshire and Staffordshire Record Offices and the National Library of Scotland. The work on the Sutherland estates was undertaken with the help of a grant in 1992 from the Royal Commission for Ancient and Historic Monuments of Scotland and I am particularly grateful to Graham Douglas of the Royal Commission for their help in this part of my work. Especial thanks must go to my husband, Peter, for his constant support throughout this project.

The following individuals and institutions have kindly allowed me to reproduce illustrations:

Royal Commission of Historic Monuments of England, 1, 23, 63
University of Cambridge Air Photographs Collection, 2, 5, 6, 7, 8, 15, 16, 17, 20, 32, 33
A.R.W. Wyman, Butser Archaeological Farm, 4
W. Rodwell, 9
Field Archaeology Division, Norfolk Museum Service, 10, 14, 25, 27, 61
St Edmundsbury Borough Council and West Stow Village Trust 11
Wharram Research Project, 12
M. Harvey, 13
Crown copyright MOD, 18
Bodleian Library, Oxford, 19
J.S. Moore, 21
C.J. Bond, 22
W. Horn and E. Born, 24
English Heritage, 28, 30
Peter Beacham, 29
Norfolk Record Office, 34
Cressingham Parochial Church Council, 35
J.C. Barringer, 37a and b
Peter Ryder, 38
Wiltshire Buildings Record, 45
Country Life, 49
Welsh Folk Museum, 50
Mrs Tom Brand, 51
Banff and Buchan District Council, 52
Historic Scotland, 53, 54
Stow Bardolph Estate, Norfolk, 60
Graham Douglas, Royal Commission on Ancient and Historic Monuments of Scotland, 67
Countryside Commission, 68, 69, 70
Cambridgeshire Record Office, 73

Introduction

Farms and fields are at the very basis of our existence. It was not until man began to till the soil that he was able to give up his nomadic existence. The 'Neolithic revolution' marks the beginning of a settled way of life and with it, the laying out of fields and their associated farms. In some parts of the country, particularly the more pastoral ones, fields that were laid out in the Bronze Age are still in use today. Only the farmstead has moved to a new site. In many places pre-Roman fields have been remarkably resilient to change, being subsumed in strip or furlong boundaries during the period of open-field farming in the early Middle Ages and then re-emerging as field boundaries with the enclosure of the open fields from the sixteenth and seventeenth century. It is only since 1950 that 2,000 or more years of continuity have finally been broken as hedges have been ripped up to satisfy the demands of modern agriculture.

Fortunately there are two sources that enable us to reconstruct this historic landscape. First there are the RAF vertical air photographs, taken in 1946 and 1947 and covering the whole country, at a scale of 1 inch to the mile. Not only do they show the countryside as it was at the end of the Second World War and just before the destruction of hedges began, but also they show the underlying earthworks of previous settlements and farming systems, many of which were flattened by ploughing and land improvement in subsequent years. We can see the British countryside in the 1940s, with far more meadow and hedgerow than now. Even the heaps of farm yard manure dumped around the fields ready for spreading can be seen in these snapshots of farming in progress. But there are other visible signs of the past: old hedgerow lines and ditches, moats, the ridge and furrow of the open fields and the earthworks of long abandoned farms and villages, all show up clearly from the air in a way that cannot be seen from the ground. In many cases, these photographs are their only surviving records.

Secondly there are the first edition 6 inch Ordnance Survey maps, surveyed between 1860 and 1890. Not only do they mark all field boundaries at a time when they were at their maximum extent, but also all the trees along that boundary. The 6 inch maps were indeed a remarkable surveying achievement and in the field boundaries the history of the farming landscape is displayed. Small, irregular stone-walled, or well-treed hedges, indicate the oldest fields of a settlement, whilst alongside these, in the areas where open-fields or commons survived until the era of parliamentary enclosure, there may be large regular fields, usually with no standing timber along their quick-set hedges. Other boundaries may also show up in the hedges on these early maps. For instance, they may respect the lines of a ring fence, denoting the edge of an earlier estate or park, perhaps a medieval deer enclosure. The pattern of small, elongated, rectangular fields running parallel to each other may indicate a Roman or earlier layout, whilst some fields still retained the long, thin and slightly curved boundaries of the medieval strips.

Maps, whether old or new, are the major tool of the landscape historian, and sites marked on the modern Ordnance Survey maps are frequently referred to in this book. They are located by means of a *grid reference* (GR), based on the national grid which allows a place to be pinpointed within a kilometre gridsquare by a unique combination of two letters and six or eight numbers. How the national grid works is explained in the margin of any *Landranger* 1:50,000 Ordnance Survey map.

Unlike the fields in which they stand, farmsteads have changed location several times throughout their history. A few are still on their Domesday sites, especially some of the isolated 'vills' of the upland regions. Others date from the Middle Ages, standing perhaps beside the church, within a moat or on the village street. Others are a result of the new creations of the time of population growth before 1400, when new land on the edge of settlements, or in woodland was brought into cultivation. Still more date from the reorganisation of farm land in the eighteenth and nineteenth centuries. These stand, often isolated in the middle of their regularly laid out fields. The earlier farms, associated with the communal open fields were centrally placed, within a nucleated village, as indeed they still are in England's most complete surviving example of an open-field village, at Laxton in Nottinghamshire.

More recently, farm amalgamations of the last thirty years have left some farmsteads redundant and replaced by huge modern prefabricated buildings on their concrete aprons at one centrally placed depot.

As well as the distribution of these farmsteads, their component buildings are of importance to landscape history. The earliest surviving buildings are usually barns and these include some of our finest examples of medieval architecture. Sometimes there was a stable incorporated into the barn and by the sixteenth century free-standing stables were being built, some of which still stand. A few granaries and cowsheds survive from the seventeenth century, but the great period for rebuilding farms was from the early eighteenth century and the majority of farm buildings date from the last 250 years.

From the eighteenth century, the influence of the great estate was at its height and landlords were responsible for rebuilding whole farmsteads, often on a carefully thought-out plan. These 'model' farms are to be found across Britain as this wave of enthusiasm for 'improvement' swept the country, reaching its height during the Napoleonic Wars. Some of our most impressive buildings date from the era of 'high farming' in the mid-nineteenth century, when industrial-style farms, more reminiscent of railway than rural architecture, were erected across the country, using mass-produced cast-iron fittings, bricks and the ubiquitous Welsh slate. This was the last great age of farm building before the depression of the 1870s which lasted, on and off, until the 1950s and the tractor era.

However, this sort of building was only suitable on farms of over about 200 acres. In parts of the country where smaller farms remained the typical holding until recently, particularly on upland livestock farms, a simpler, vernacular style of building lasted much longer and contributes to the great variety of our farming landscape. Despite the wholesale destruction and standardization of the last thirty years, the British countryside is still a fascinating landscape to behold and study. It is the aim of this book to help the reader understand what he sees and to appreciate what this contributes to our knowledge of the agricultural history of the past 5,000 years.

THE EVOLVING LANDSCAPE

1 The beginnings of agriculture - the earliest fields

The first cultivators

Man has been cultivating the land of Britain for at least 6,000 years and in some places that cultivation has been almost continuous. In such circumstances one might expect that the chances of any evidence of those early farmers surviving would be very slim indeed, and yet, even in the most intensely worked areas of the country something still remains. Elsewhere, in regions which have since become marginal land, there is more to be found.

We know that the first Neolithic farmers were burning and felling the forests with their superior axes made from hard rocks such as flint by about 4000 BC. They were growing cereals and keeping cattle, sheep, goats and pigs. By 3000 BC they had a simple plough, known as an ard, which they pushed through the ground, but very little remains of the settlements of these people. Many were probably not occupied for more than a few years. When the land lost its fertility, they would move on, destroying more forest. As the population grew, from about 30,000 to 50,000 between 5000 and 2000 BC,[1] deforestation continued to gather pace and more land was needed for cultivation. By 100 BC, there may well have been less woodland in England than there is now,[2] although in Scotland, the Great Caledonian forest survived, almost intact, until the end of the Middle Ages.

The great religious and funereal monuments of the Neolithic Age, coupled with the evidence for axe factories trading their products over great distances, indicate the existence of a complex society. The distribution of finds shows that the countryside was already densely populated and its remotest corners had been penetrated. A fairly sophisticated type of agriculture, based usually on the single farmstead, but occasionally in larger, village-style settlements, was necessary to support this society; one in which perishable goods such as hides, cereals and animals must have figured as items of exchange, as well as the axes and pottery which survive as evidence for a highly developed trading community.

This agriculture must have involved the laying out of fields and perhaps even the demarcation of territorial units, but it is not until the period starting about 2000 BC (after the introduction of the first metal tools, made of bronze) that any evidence for fields can be found.

Bronze Age fields

Complex field systems dating from the second millennium BC have been identified from air photographs in an area now mostly covered by the Fengate industrial estate outside Peterborough. Connected with this system of early ditched enclosures are droveways. This, coupled with the fact that the entrances to most of the enclosures are in the corners (a position convenient for the

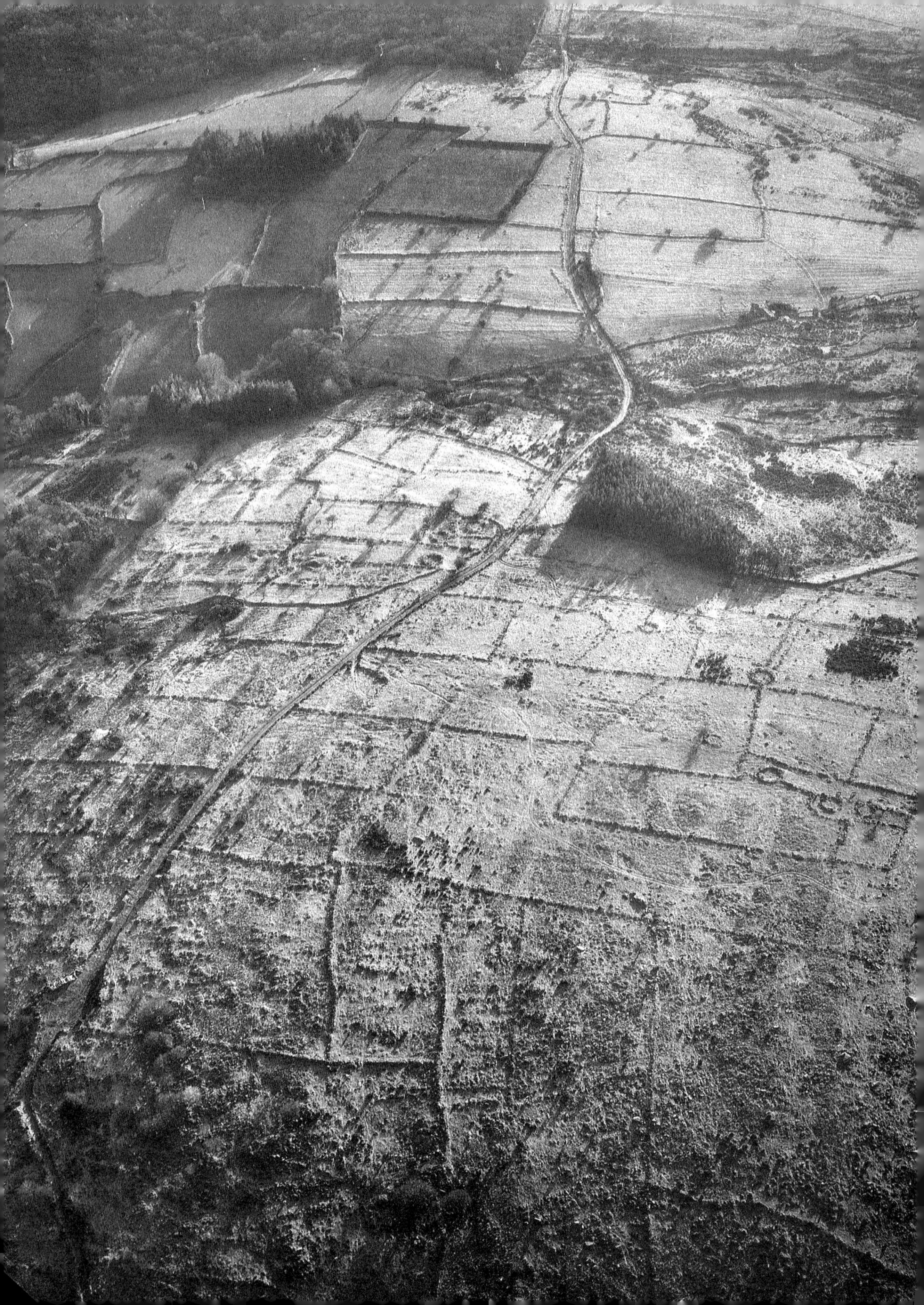

The beginnings of agriculture – the earliest fields

1 *Area of Bronze Age fields on Dartmoor (GR SX7574) associated with Bronze Age circular huts (RCHME)*

driving of livestock) suggests that they were designed for stock. The ditches were for drainage to provide flood-free winter grazing. An impressive attempt to reconstruct these Bronze Age fields has been made at nearby Flag Fen where the Fenland Archaeological Trust is creating a Bronze Age landscape with fields, droveways and natural vegetation as well as round huts and farm animals.[3]

Other examples of ditched enclosures dating from the Bronze Age have come to light in other parts of eastern England, whilst the most impressive walled field systems, also used for enclosing grazing stock are to be found on Dartmoor.

These walled enclosures, containing Bronze Age farms, were created over a short period between 1700 and 1600 BC and remained in use for about 300 years; some, indeed, are still used today. They cover over 2,000 acres (800 hectares) of the Dartmoor landscape between 1,000 and 1,400 feet (300 and 425 metres) above sea level. Walls known as reaves, about one metre wide and still standing 20–30 inches (50–80 centimetres) high, run in parallel lines about 300 feet (900 metres) apart for miles across the countryside. The main reaves are divided by cross-reaves at irregular intervals. The system peters out when it reaches the steeply-wooded slopes of the Dart valley, but then carries on, on the same axis, up the other side. They are best seen on Holme Moor (SX6771) by looking north from Combestone Tor on the road between Holme and Hexworthy.[4]

2 *Aerial photograph showing the Bronze Age ditches at Fengate, Cambridgeshire. The long parallel ditches mark the edges of the drove road, and the small rectangular patterns, the fields off it (University of Cambridge Air Photograph Collection)*

The size of these field layouts suggests a degree of planning on a gigantic scale by communities able to parcel out tens of square miles as they pleased, rather than by smaller groups. They are one example of the great replanning of the countryside which took place in the Bronze Age and had an impact on the landscape comparable to that of the late eighteenth century parliamentary enclosures. It is tempting to suggest that there must have been some sort of central authority, but whatever the method by which it was created, the result was a landscape of scattered farmsteads amongst planned fields. Each Dartmoor territory has access to high moor, probably common grazing, again an indication of a high degree of central planning.

These boundaries were often ignored by later fields, and the medieval walls, taller, and often with ditches beside them, should not be confused with the earlier ones. That the early fields survive at over 1,000 feet (300 metres) is an indication of the intensive farming practised in the region in the Bronze Age. No doubt fields existed at a lower level as well, but there they have been superseded by later enclosures.

On the downs of Wessex ditched and banked enclosures of the late Bronze Age, again covering huge acreages, have been identified and interpreted as cattle kraals. They are of various shapes and sizes from a half to two acres. Both sides of the Bourne valley are scored with a complex system of V-shaped ditches running down to the river and enclosing large tracts of land, sometimes cutting across earlier field systems.[5]

Walled systems from the Bronze Age rarely survive elsewhere, but on the Land's End peninsula 'there is one of the most impressively ancient farmland landscapes in Europe'. Bronze Age objects buried in some of the banks of the field boundaries in West Penwith are an indication of the age of the tiny irregular fields which form one of the 'world's oldest landscapes still in use.'[6]

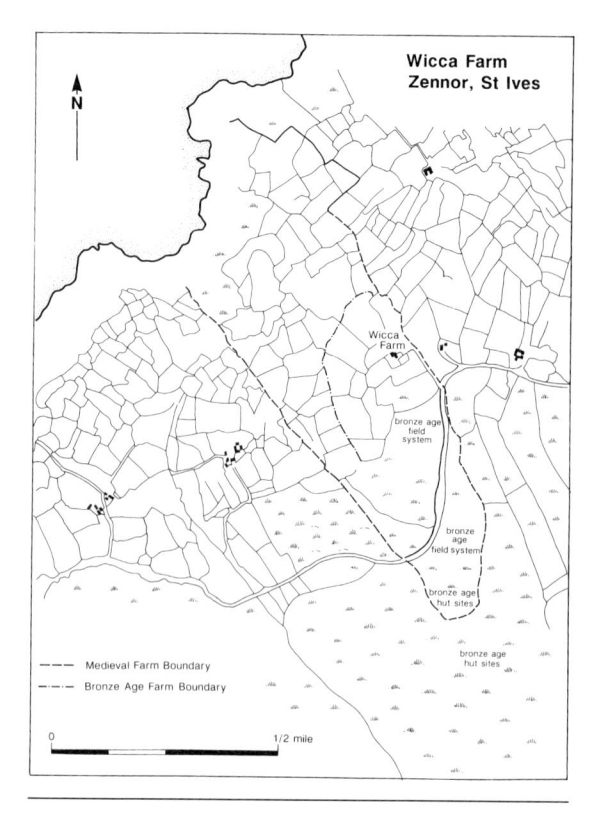

3 *The site of Wicca Farm, Zennor, Cornwall has been continually occupied since the Bronze Age. The boundary of the original farm can be seen in the curved wall around the farm. The stone banks, surmounted by hedges enclose tiny fields, all of which were in existence by the Middle Ages (after Nankervis, J. The traditional farm, Wicca, Zennor, St Ives, MAFF 1989)*

Similarly Bronze Age hut circles occur within the field systems at Zennor and the present farm of Wicca included the great semi-circular bank still standing to 10 feet (3 metres) and enclosing 25 acres (10 hectares). Four small fields near the house were manured and cultivated whilst five larger units further away were grazed and cultivated in turn.[5] The characteristic Cornish 'hedge', or stone-faced bank, the product of the necessity to clear a field of stones as well as the need for shelter from the salt-laden winds, have in many cases, been built and maintained continuously for 5,000 years.

Remains of reave-type systems have also been identified in Nottinghamshire and Bedfordshire, and no doubt others await discovery as the relationship between prehistoric finds and surviving field boundaries are established.

Other field patterns have been located as soil marks on air photographs and it is becoming clear that much of England, from Sussex to Northumbria, was cultivated in fields by the end of the Bronze Age.

By the end of the Bronze Age, the inhabitants were finding it necessary to build defended hill-fort type settlements alongside their planned fields and isolated farms within a well-defined territorial system – all of which suggests competition for land leading to warfare within a landscape already highly populated, with a population perhaps approaching one million.[8]

Many of the Bronze Age fields are associated with pastoral farming, with cattle being the most important animal. Pigs, sheep and goats were of less significance, along with hunting, particularly for red deer. Arable farming was also practised, even at high levels. Evidence from peat bores on the North York Moors shows that barley and two primitive types of wheat, emner and spelt were grown. By the Iron Age these uplands had lost their fertility as a result of early farming activity and over-exploitation. Arable fields became confined to the lower land.

The introduction of iron

Iron was introduced into Britain about 800 BC and with it came a more efficient plough. An ever-increasing population meant the cultivation of ever more land. Most Iron-Age fields were used mainly for arable. From pollen analysis, seed impressions on pottery, carbonized seeds found in excavations, and the investigation of food remains in the stomachs of bodies preserved in peat bogs, we can learn about the diet and therefore the crops grown by Iron-Age farmers. Emner, spelt, barley and oats were grown, barley being the most important. Barley is easier to thresh out than primitive wheats and was used for flour and brewing. Field beans, vetches and fat hen (now regarded as a weed) were also grown and it is possible that the beans were grown in some sort of rotation, thus helping to maintain soil fertility. More land was ploughed up and more woodland cleared. A variety of farmsteads, from isolated to larger groupings, were set up and by the Roman period, Britain could be described as an over-exploited country. It was indeed Britain's reputation as a corn producer which was one of the factors that influenced the Roman decision to invade.

Whilst nothing of the Iron-Age farmsteads themselves remains in the landscape, they have been located through excavation and attempts to recreate Iron-Age farms from the archaeological evidence have been made. One such site was at Butser Hill near Petersfield in Hampshire. On a northern spur of the chalk downs an Iron-Age farming landscape was reconstructed. Huge round huts built either of stone or wattle, and with thatched roofs were built as workshops and dwellings, while smaller huts which would have been for animals and storage were also erected. The farmstead was surrounded by a bank and ditch, hardly large enough to be defensive, but perhaps providing some protection from wild animals. Half acre (0.2 hectare) fields were created. This is the size suggested from nearby field boundaries, located from aerial photographs, and it also happens to be the area that one man and an ox-team could plough in a day. Fields for animals were surrounded by wattle fences to keep stock in, and where possible the animals were of primitive breeds, as near as possible to the types that would have been kept in the Iron Age.[9] The farm has now been moved to Charlton, still in Hampshire, on the Downs, just off the A3. In a quiet valley, pre-historic

4 Reconstructed Iron-Age huts beside their fields enclosed by wattle fencing at Butser Hill, Hampshire (A.R.W. Wyman, Butser Archaeological Farm)

fields are to be seen to the north, and a fine piece of ancient woodland to the south. Here in this open-air archaeological laboratory, experiments both in animal and plant breeding take place. For instance, attempts are being made to produce a white Soay (a brown Iron-Age type sheep) and also to trace chickens back to their beginnings as a domesticated fowl, probably in the Roman period. The primitive wheats, emner and spelt are grown, and different types of ards are used for ploughing using the medium-legged Dexter cattle. The site is open to the public who can see how prehistoric farming worked in practice.

Other 'Iron-Age farms' are at Cranborne in Dorset, on Donnington Heath, Leicestershire, run by the Leicestershire Museum Service and on the Pembrokeshire coast in the National Park. 1992 marked the centenary of the discovery of Glastonbury Lake Village and there are plans to reconstruct part of the village within its environment.

Evidence for Iron-Age fields is more widespread than for earlier periods. Where land was ploughed on sloping ground, small banks grew on the downhill boundary. These banks are called lynchets and are formed by the steady, slow movement of soil downhill. Hill sides of lynchets, found particularly on the chalks of southern England, are thrown into sharp relief by the long shadows of morning or evening sun and show how far up onto high ground the Iron-Age farmers were penetrating. Other field systems can be seen from the air as crop marks, sometimes in totally contrasting areas such as the river gravels of the upper Thames valley, and on the chalk downs of Wiltshire.

Finally, and perhaps most interesting because of the continuity of agriculture which they demonstate, are those Iron-Age field patterns which have survived as field

5   Strip lynchets, Askerswell, Dorset (University of Cambridge Air Photograph Collection)

6   'Celtic' fields, Fyfield Down, Wiltshire (University of Cambridge Air Photograph Collection)

boundaries into this century. Because fertile areas such as East Anglia have been farmed without a break, does not mean that nothing remains of earlier field systems; in fact the situation is quite the reverse. In parts of Essex, Suffolk and Norfolk, rectilinear hedge patterns have survived into modern times.[10] In a large area of south-east Norfolk between the rivers Chet and Waveney field alignments are similar, suggesting they were laid out as a block. As they are crossed by a Roman road, they must date from a pre-Roman period.[11] In the neighbouring Dickleborough area of south Norfolk the hedges, as shown on the tithe surveys of the 1840s, formed a rectilinear pattern which was clearly cut across by the Roman road, the modern A140, and so must be older than it. This field system could be traced over a large area covering at least a 5 mile (8 kilometre) square. Similar systems survive over an extensive area of central and north Essex which continue to form the basis of the present landscape, and there is no doubt that others await discovery. Many field patterns are probably very much older than has previously been thought and we are gradually coming to understand the extent of the pre-medieval farming landscape that survives.[12]

The Roman impact

The invasion and conquest of southern Britain after AD 43 involved the suppression, by about 40,000 men, of a country inhabited by about two million people.[13] Most of Scotland was never subdued and here pre-Roman-style buildings and farms continued to be built. Farmsteads continued to be circular in shape and of timber-frame construction. At Scotstarvit in Fife, three successive timber-framed round houses were built within a ditched enclosure. Another circular house with a low stone and turf wall surmounted by a timber superstructure, this time in a rectangular enclosure, has been excavated at Greencraig, also in Fife. At West Plean in Stirlingshire two successive circular houses appear to have been occupied from the late Bronze Age to the first century without a break.[14]

We are also only now appreciating the density of pre-Roman population. The land was already divided into fields and there was little scope for the Romans to create a new landscape. They took over the existing fields and frequently farming and farmsteads continued much as before.

However, the Romans did introduce their own distinctive type of fields. They were longer than they were wide, in a ratio of up to five to one. They were grouped together and laid out in blocks of parallel fields. These blocks can be seen in many areas, such as on the chalk of Wessex and Sussex, on the limestone of Wharfedale, Yorkshire, and in Somerset, and the Fens.

Two levels of farming dominated Roman Britain. First, the peasant farmers continued much as before in either nucleated settlements or isolated farms, cultivating the irregular Celtic fields, living in round huts and storing their grain in storage pits. Here, the veneer of Roman life was thin. A few of the cheaper mass-produced luxuries of Roman civilisation, such as pottery and trinkets, might reach the farms and gradually, as at Park Brow in Sussex and Studland in Dorset, the Roman ways began to take over with rectangular houses, granaries and grain-drying kilns finding their way onto native farms.[15]

Secondly, there were the villas. These houses with their tiled roofs and tessallated floors, represent a higher level of culture and comfort. They grew up beside and out of the old, exploiting the new opportunities for commerce provided by the towns, roads and resident garrisons that needed feeding. Most villas were near roads and not too far from towns. Many, such as Chedworth in Gloucestershire, concentrated on large-scale livestock farming, which was more profitable than arable. Their prosperity increased in the third and fourth centuries. Many villas were reconstructed Iron-Age farms and were enlarged several times.

In some areas, the countryside was already over-exploited, but the population continued to rise and farming to intensify during the 400 years of the Roman occupation. New and better ploughs, with a coulter fitted in front of the share, gradually spread from the romanized villa-type agriculture to the local farms and allowed for more land to be cultivated. New roads and towns made it easier for farmers to trade their grain and a foreign market too was opened up. The Romans built granaries for the storage of grain and these replaced the traditional Iron-Age grain-storage pits on farms. There was a military market for food which encouraged cereal production in the uplands of the north. At Grassington in Upper Wharfedale, North Yorkshire, lynchets cut into the hillside are evidence for arable farming over 2,500 acres (1,000 hectares of upland).[16] There were many villas in East Yorkshire by the fourth

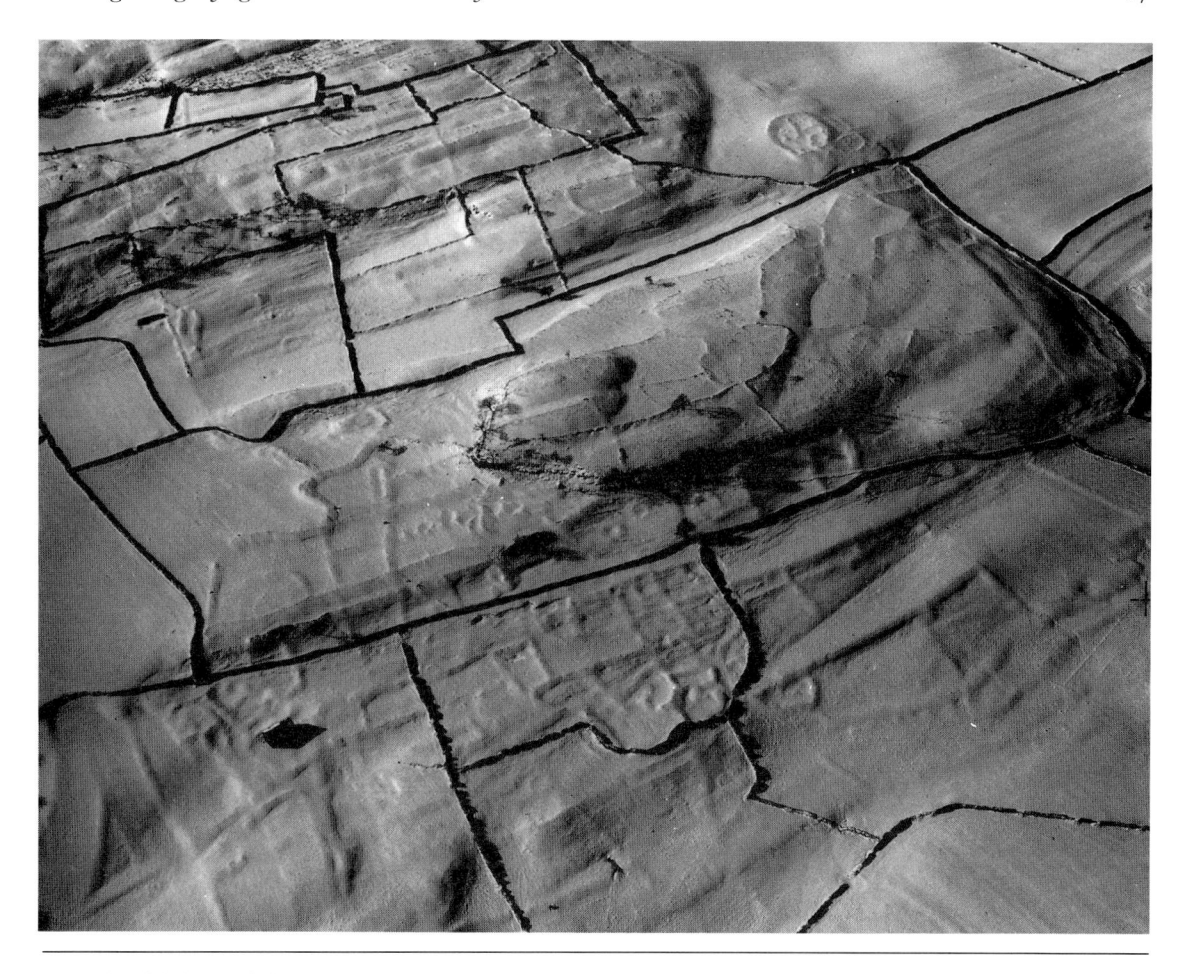

The beginnings of agriculture – the earliest fields

7 *Wharfedale, Yorkshire showing the walled enclosures underlying the medieval field boundaries, mostly dating from the Roman period (University of Cambridge Air Photograph Collection)*

century, as well as in the Pennines and Cumbria. A corn tax meant all farmers had to hand over part of their crop to the army. New crops such as rye, flax, apples, cherries, cabbage and carrots were introduced. A stable government provided conditions in which agriculture could prosper.

Perhaps it is not surprising therefore that there was a population explosion at this time of increased agricultural production. Many field systems are older then the Romano-British farmsteads placed within them, and presumably larger farms were being split up to cater for the increased population which probably reached four million,[17] a level higher than it was at the time of Domesday Book. In east Northamptonshire there were at least two farmsteads for every square kilometre, and these sorts of figures may well be applicable elsewhere.[18]

There was a great variety of farmsteads, from isolated farms and villas to hamlets. One such hamlet was excavated at Studland in Dorset, dating from the beginning of the Roman period in the first century. It began with a group of wattle and daub, circular, Iron-Age-type houses. After less than a generation these were replaced by rectangular houses of a 'longhouse' type with two rooms, one for animals and the other for humans.[19]

At Stanwick, not far from Raunds in the Nene Valley area of Northamptonshire, the site of a Roman villa has been known since the eighteenth century, but recent work has shown that alongside the villa is a planned series of ditched enclosures linked by trackways, covering at least 70 acres (28 hectares) and laid out in the second century. Originally these buildings were of a circular Iron-Age type, but gradually they were replaced by rectangular buildings within walled yards. There is an Iron-Age settlement nearby, so it is clear that the Romans were not taking over an empty landscape, and that the change from Iron-Age to Roman-type layouts was slow.[20]

A very different stone-built hamlet survives high up on the moors of Cumbria at Crosby Ravensworth. Its position, over 1,000 feet (300 metres) above sea level, suggests that pressure of population meant marginal land was being cultivated in areas not ploughed since. As a result the stone-walled house sites, yards and paddocks remain, undisturbed by later activity. Similarly at Crave, North Yorkshire, there are stone-walled fields associated with circular and rectangular houses.[21]

The villa is the type of Roman rural settlement with which we are most familiar, yet many of these were little larger than family farms, probably employing mainly family labour. They were the centre of Roman-style estates which varied greatly in size. The boundary ditch of that at Ditchley in Oxfordshire enclosed 875 acres (354 hectares) and it is now followed by parish boundaries.

Although evidence for Roman farms at the centre of their fields or estates comes mainly from excavation, some estate boundaries, such as those at Ditchley, remain fossilized in later boundaries, or fields in modern landscapes. In Essex and north Kent there is a huge area of planned Roman fields stretching from Thurrock on the north side of the Thames, across the river into north Kent. Surviving as roads, tracks, parish boundaries and crop marks, this field pattern appears to be evidence for a huge imperial estate.[22]

8 *Roman settlement at Crosby Ravensworth, Cumbria. The stone walls of the fields and paddocks around the circular houses and farm buildings are clearly visible on this upland and isolated site (University of Cambridge Air Photograph Collection)*

Whilst farmsteads have frequently moved around within their land holdings through history, the boundaries of some of these estates have remained remarkably constant. 'Whatever happened to the settlements, the land units would tend to be more stable for they were the basis of life'.[23] As at Ditchley, Roman estate boundaries, which in some

The beginnings of agriculture – the earliest fields

cases go back to the Iron Age, may survive in the form of parish boundaries. Similarly, land that in 1086 was held by the Bishop of Chester in a simple block around Lichfield was probably the land associated with the small Roman settlement of Lekocetum, just south of Lichfield.[24]

As much of Iron-Age Britain was already cultivated when the Romans arrived, there was little scope for the complete relaying out of fields except on new lands such as those emerging from the sea, along the Bristol Channel and the Wash. In Somerset and South Wales there was large-scale drainage with regularly spaced ditches creating rectangular fields. In the Fens around the Wash Roman engineers constructed a complex system of drainage ditches and reclaimed large tracts of land. The Car Dyke ran around the fens from near Cambridge to Lincoln, and the Foss Dyke from Lincoln to the Trent. Both these canals were for transport as well as drainage, ensuring communication links between this isolated area and the main military markets for agricultural produce. Since this was new land with no previous owner, it became the personal property of the emperor and a huge imperial estate was established from Denver on the Norfolk fen edge, through to Spalding and further north, almost as far as Lincoln. The extensive Roman site at Stonea Grange (Cambridgeshire) may well have been the administrative centre for the area. It

9 *A huge area of rectilinear Roman fields can be traced across the area north of the Thames around Thurrock, Essex. Roman sites are stippled (W. Rodwell)*

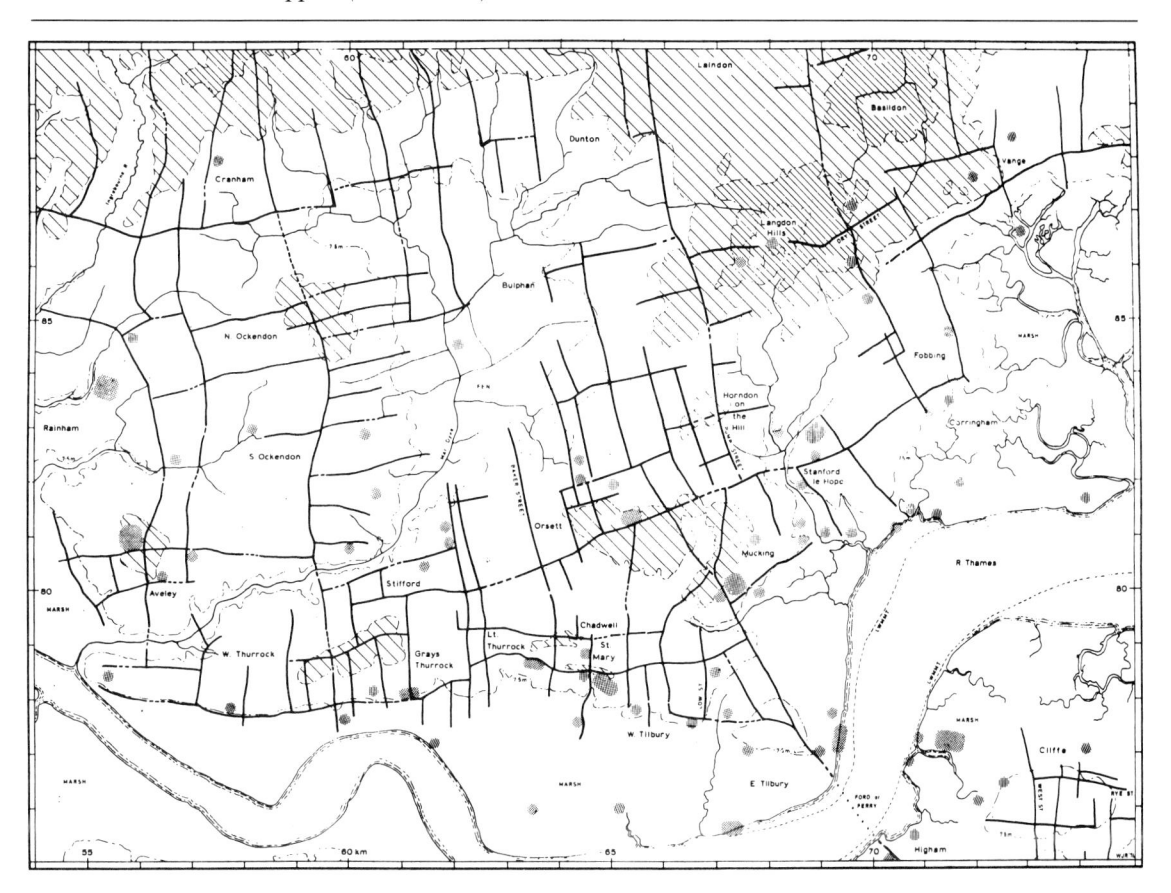

10 *The fenland was a densely populated and intensively farmed area in Roman times. This photograph shows the Roman road, the Fen Causeway, as a white sinuous line following an old river bed, or roddon, across the centre of the picture. Below it are the rectangular fields of a highly organized Roman farm (Field Archaeology Division, Norfolk Museums Service)*

has been calculated that the land was originally parcelled out into holdings of about 110 acres (44 hectares), but that these were later subdivided. There is no continuous mesh of Roman fields across the fens, but rather small clusters displaying a great variety of shapes and sizes. These may well have been used for cereals, but more important were the large flocks of sheep and herds of cattle grazed over the fields between roadways and watercourses, and the open spaces to which the ditched roads lead. These fertile soils supported a dense population with large villages existing alongside isolated settlements. Ditched tracks enter compact areas of small ditched enclosures and paddocks. The simple, basic finds from their house sites show that the inhabitants of these imperial estates never became as wealthy as their counterparts elsewhere. Settlement in the fens came to an abrupt end as the sea level rose and the area was inundated in the fifth century.[25]

How much of the farming landscape of prehistoric and Roman Britain is still visible today? Probably far more than we realise, although again, far less than forty years ago. Farmsteads have almost all disappeared, except for those few stone-built remains, either above ground or revealed by excavations in marginal lands where they have remained undisturbed. Only at reconstructed sites such as Charlton in Hampshire and Flag Fen near Peterborough, can we begin to appreciate what the buildings and the farmed landscape may have looked like. Fields and estate boundaries, however, have been far more permanent. They can either be located as boundaries on maps or on the ground as walls, hedges and ditches. Only when the small prehistoric and Roman fields no longer suited the farming system, as was the case over parts of England in the Middle Ages, were they likely to be altered, and even then some of the original boundaries have often survived.

Summary

Very little in the way of farmsteads survives from the pre-Roman period and to find out about prehistoric farming it is best to visit one of the reconstructed farms at, for instance, Flag Fen or Charlton. As well as these set-piece sites many surviving field boundaries in the form of hedges, banks and ditches can be seen, and may be far older than they at first appear. As at Wicca, there may be prehistoric houses with them, or as on Dartmoor, they may underlie the later landscape.

Some of these boundaries, particularly in lowland Britain, may have been swept away by modern farming and so the field patterns can only be picked up on older maps, (the early editions of the Ordnance Survey 6 inch sheets are best) or on the RAF 1946 aerial photographs. Both of these sources can be consulted in the county library or archaeological unit in most areas.

Pottery scatters in ploughed fields are an indication of the site of a farmstead. Roman sites particularly are rich in distinctive pottery and the survey of fields across a few parishes will reveal the density of population in Roman times.

2 From the Romans to Domesday Book

How dark were the 'dark ages'?

The departure of the Roman armies from Britain by the fifth century was an important political event marking the end of an era in the history of the British Isles. How significant it was in the development of farming is far more difficult to ascertain. 'On the morning after the last civil servant had gone men and women still woke up to the routine of their everyday lives'.[1] The Anglo-Saxon settlers came into an already fully peopled countryside. Although, no doubt, their advance caused much death and destruction, and the lack of central control resulted in the breakdown of trade, it is quite possible that the collapse of urban life was not accompanied by a similar disruption in the countryside. There may well have been no more than 10,000 Saxon settlers, resulting in a political takeover of a disintegrating society rather than a mass replacement population. Their progress across the country was slow, not reaching the north and west until the seventh and eighth centuries.

In the western uplands the collapse of Roman rule and the slow penetration of the Saxons from the east was likely to have had only minimal significance. On the slopes of Dunkery, 1,100 feet (335 metres) above sea level, at the heart of Exmoor, the deserted farmsteads of Bagley and Sweetworthy are beside prehistoric ringworks, presumably the sites of earlier farmsteads. The farmsteads were within stone boundary walls, probably of the same date as the ringworks, which continued to form a stock-proof divide between the moor and arable throughout the Middle Ages. This provides remarkable evidence for continuity of settlement from pre-Roman times to the late Middle Ages.[2]

There are few Roman sites where a continuity of settlement into Saxon times can be convincingly demonstrated, but this is hardly surprising, partly because of a lack of archaeological evidence for permanent structures or early Saxon pottery in many parts of the country outside East Anglia, and partly because we know that settlement sites shifted frequently through the Iron Age, Roman and Saxon periods, perhaps every 150 to 200 years.

At the Iron-Age hillfort of Crow Hill, near Raunds in Northamptonshire, pottery scatters suggest that it was occupied in both the Roman and early Saxon period, which suggests that it may be the original site of the village of Irthlingborough.[3] More usually, the Saxon site is near to the Roman one, allowing for the continued exploitation of already cleared land and established fields. Gradually, over the fifth and sixth centuries, the existing pattern was adapted to suit the needs of the new community. There was certainly not a clean slate onto which the Saxons could write a new story. Much must be speculation because of the lack of landscape and other evidence. No Saxon ploughs have been found and no field systems can be dated specifically to this period so aptly known as 'the dark ages'.

Even in this well-populated countryside, marsh, fen and wetland still played an important part. Fenland covered much of Lincolnshire, Cambridgeshire, Huntingdonshire and west Norfolk, as well as the Somerset levels. The marshes of Pevensey and Romney, the mosses of southern Lancashire, Flanders and Blairdrummond in the upper Forth Valley to the west of Stirling, the Howe of Fife and the Carse of Gowrie all remained ill-drained until at least the seventeenth century. Marshland also surrounded many river estuaries from the Thames to the Dee and fringed the coast from Hampshire to the Solway Firth. These areas were not, however, desolate wildernesses, but supported a complex and sophisticated way of life based on wild fowling, fishing, the cutting of reeds and rushes, and the exploitation of rich summer grazings. Reclamation in many of these areas had been started by the Romans and continued throughout Saxon and early medieval times, only to be destroyed in the floods of the early fourteenth century.

Woodlands too were carefully parcelled out and utilized. Their extent and boundaries were always described in detail in Saxon charters. Coppicing, pannage for pigs, wood for building and fuel were all valuable.[4]

It is clear that the medieval agriculture of much of lowland Britain, based upon open fields divided into strips and worked communally from nucleated villages, did not originate in this initial period of Saxon settlement. The few sites that have been excavated are small, consisting of a few rectangular hall-type houses alongside smaller huts, used as workshops or store houses.

The 'village' of West Stow in Suffolk has been fully excavated and the evidence for its agriculture studied. Some of the buildings have now been reconstructed and form part of a country park administered by St Edmundsbury District Council. Not far away was a Roman villa and it is possible that the incoming Saxon settlers found an unused corner of heathland in an otherwise flourishing estate and built their houses there, perhaps protecting the estate from later arrivals. As the Roman system crumbled, the Saxons took control of the farmland. This small, isolated settlement probably consisted of no more than three or four families at any one time, each family occupying a hall, outbuildings and sheds. There is no evidence for either Roman or Saxon fields, but these were probably worked as separate units with an 'infield' area nearest the settlement farmed intensively, and a more distant 'outfield' used mainly for grazing and only ploughed occasionally. Such a system was perfectly adequate so long as population pressure was not great.

Animals played an important part in this mixed farming economy. Sheep were the most numerous, kept for their wool and milk as well as meat. Bone evidence shows that forty per cent of animals were slaughtered at under a year, suggesting the importance of milk production; nearly forty per cent being slaughtered between one and four years, suggesting meat production; and twenty per cent were slaughtered at over four, suggesting wool production. Sheep were therefore valued for all their potential uses.

Although sheep were the most bred animal, cattle would have provided the most meat, being mainly slaughtered between two and a half and four years of age. However, as the majority of the older examples are female, dairying must also have been significant. The importance of pigs gradually declined during the fifth century, to be replaced by cattle. Goats and horses were also kept in small numbers. The types of animals kept, their size and the ages at which they were slaughtered have more parallels with the preceeding Iron-age and Roman farms than with the Saxon farms of Europe, suggesting that there was indeed considerable continuity from the Roman period, and perhaps we are looking at the settlement of some 'Saxonised Britons'.[5]

If there was continuity between the Roman and Saxon period as far as animals were concerned, this was not so true of crops. Material analysed from archaeological excavations shows that the range of cereals cultivated seems to have changed abruptly. The old spelt and emner strains of wheat and six-row barley were replaced before the eighth century by those nearer their modern equivalents such as bread wheat and two-row barley. Oats and rye are found for the first time. In most regions suited to arable farming the entire range of cereals were now grown, but in varying proportions. Perhaps not surprisingly, rye was grown on light soils. In southern England wheat, followed in importance by barley and oats, were the main cereals, with very little rye. Large quantities of oats were produced, mainly for animal feed. Hemp and flax were also of local importance.

The village at West Stow did not survive more than 200 years and it is clear that the hallmark of early Saxon settlement was its mobility. It is perhaps because of this mobility that there is nothing in the farming landscape today that can be specifically dated to this period. We can only guess at the field and farming systems that were practised.

At Raunds in Northamptonshire, for instance, excavations have shown that settlement in the sixth and seventh centuries was an irregularly organized loose agglomeration, perhaps extending over 25 acres (10 hectares), but in the late seventh century, there was a dramatic replanning and contraction of settlement into a rectangular ditched enclosure.[6]

The emergence of the nucleated village

It was in the eighth and ninth centuries, perhaps associated with the Scandinavian invasions rather than at the end of the Roman Empire, that the major changes in farming history which were to affect the landscape of Britain took place. This is the period when the many early dispersed

11a and b *Reconstruction of part of the Anglo-Saxon village at West Stow, Suffolk, and a plan of the fields around the settlement as they might have been at the time of the conversion to Christianity, c. AD 650 (St Edmundsbury Borough Council and West Stow Anglo-Saxon Village Trust)*

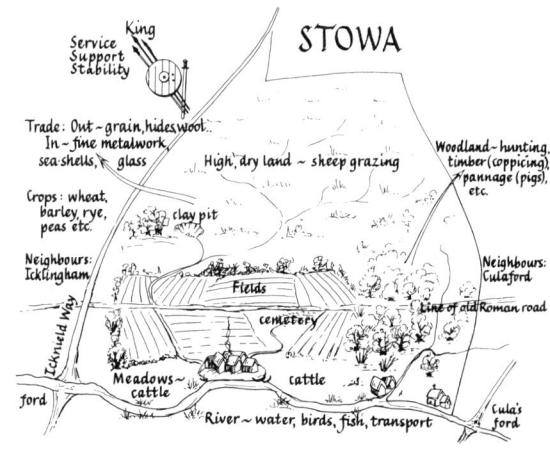

settlements, often only indicated now by pottery-scatters in the fields, began to be abandoned, and nucleated settlements established. Rarely do excavations on deserted medieval village sites provide evidence for pre-Christian settlement, and, similarly, neither does field-walking in the area around medieval churches produce early-Saxon pottery scatters. This does not mean that the area was either wasteland or unoccupied earlier; rather that the sites of farmsteads have moved. At Raunds, replanning in the later ninth and early tenth century involved the laying out of at least three rows of holdings, coinciding with the building of the first church and a large timber hall on the site of Furnells Manor. Perhaps there was a standard unit of measure which was employed to create identically sized plots within each row – this could be the manorial system of medieval England in an early stage of creation.[7]

In the six Wharram townships in the Yorkshire Wolds there are no signs of early Saxon settlement and little late Saxon pottery. All the villages have a planned layout which was either created in the Scandinavian era, or more probably in the twelfth century. The creation of the villages must coincide with the laying out of the open fields which can be reconstructed from aerial photographs. The field system shows a remarkable regularity. Strips stretch for up to 3,200 feet (1,000 metres), ignoring the general topography. The old Iron-Age landscape was thus completely changed, the scattered settlements being overlaid by the new field system, and the winding tracks replaced by regular access headlands between the furlongs. This large-scale replanning, which covers the whole area of the Wolds plateau, must have been carried out at a time when population pressure necessitated a full exploitation of the soil, probably at the time of maximum population in the twelfth century.[8]

Field-walking to record scatters of medieval pottery in the ploughsoil can indicate not only where houses were, but also which areas of the fields were most heavily manured. Bits of broken pottery would be spread on the fields with the household rubbish, compost and the contents of cess pits. This type of study has shown that only certain areas near the village were manured, suggesting an intensively cultivated infield, with the outfield less used. Many of the well-manured areas correspond with those used in Romano-British times, suggesting continuity of both farming methods and sites. The Romano-British estates, which may well have been laid out in the Iron Age, could have been the basis of later townships, providing continuity of boundaries as well. An older set of boundaries was respected when the open-fields were laid out as there is a break in the strip pattern at the township boundaries which may well respect those of the previous Romano-British estates.

Although Somerset has relatively few large nucleated villages set in the middle of open fields, there was a degree of nucleation and replanning of settlements some time between the tenth and twelfth centuries. This was possibly as a result of the manorialization of post-Norman England – a change that is likely to have been of greater significance in the south-west than either the Saxon or Scandinavian incursions.[9]

In the much-studied parish of North Elmham in Norfolk, settlement near the huge cremation cemetary of Spong Hill was abandoned before the mid-eighth century in favour of the present village site a few miles to the north. 'Some force was strong enough to obliterate the old settlement . . . whether it was Christianity, working through the early bishops, or the will of some great lay landowner, or a social impulse, or an overiding agrarian reason, may never be known'.[10]

The more sites that are studied the clearer it becomes that in lowland Britain there was

0   150   300 feet
0   50   100 metres

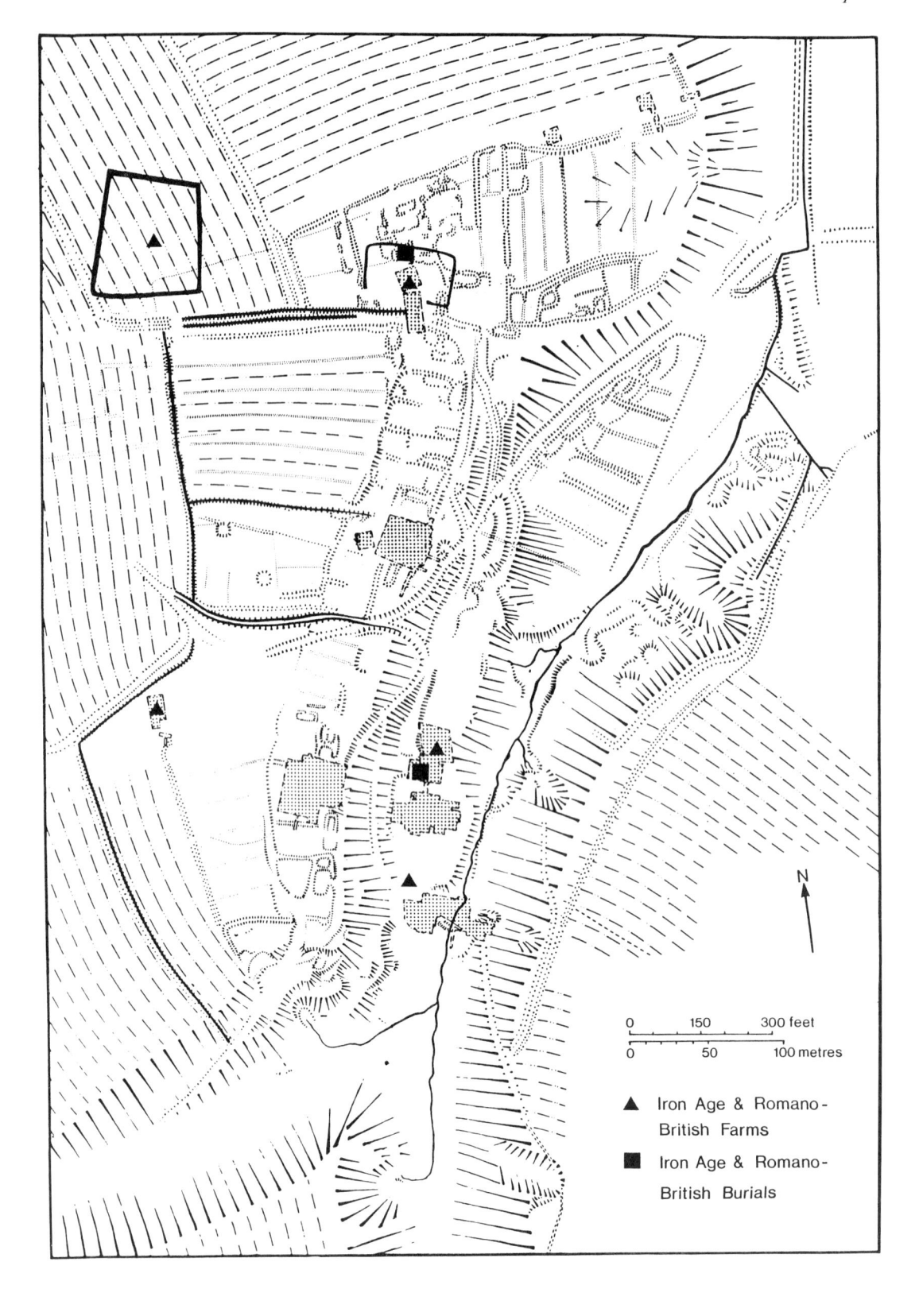

▲ Iron Age & Romano-British Farms

■ Iron Age & Romano-British Burials

12 *The village of Wharram Percy, Yorkshire, showing the Iron-Age and Romano-British sites and the way their boundaries have been respected by the later settlers. The older boundaries have been marked in heavy lines (Wharram Research Project)*

a trend somewhere between the seventh and ninth centuries for isolated sites and hamlets to be abandoned in favour of nucleated settlements. What was it that persuaded the inhabitants to make this move which must have been accompanied by a change in the farming system? This is a fundamental question to which it is impossible to give a conclusive answer, but it may be the result of population pressure, perhaps intensified by the arrival of Scandinavian settlers. This might explain the Scandinavian names of many east coast villages, perhaps re-sited during this period of major dislocation in the area covered by the Danelaw. It is the most likely time for the laying out of the village at Wharram Percy, and perhaps parts of Holderness in east Yorkshire. Here areas much larger than the parish have been planned with such extraordinary regularity that it would appear that there had been some central authority. The only time when the area was not split amongst a variety of landowners was during the years of Danish control in the late ninth and early tenth centuries.[11]

In areas of East Anglia where field-walking has been carried out, the evidence from pottery-scatters certainly indicates a rapid rise in population by the tenth century. Small farms became villages and new sites were occupied. This argument is strengthened by the fact that in the north and west of England, where population pressure was far less, isolated farms and hamlets, working a compact group of fields within a ring fence, have remained the typical form of settlement. However, it is also true that there are areas of north Essex and Hertfordshire where population pressure was as great as in neighbouring districts, but where small hamlets and ancient farmsteads with early Saxon, Roman or earlier origins, survive. The explanation for the variety of farming types indicated by the landscape evidence is therefore far from simple.

There is no single answer to the many anomalies in settlement patterns and the distribution of isolated and nucleated settlements and the type of farming they supported. A powerful landlord did not automatically mean nucleation of a settlement. We have to accept the fact that even in neighbouring manors medieval people lived under differing rules and traditions, with different ways of organizing their farming activities.[12]

Some indication of the degree to which Anglo-Saxon England was exploited for agriculture is given in the Domesday Book.

13 *Reconstruction of the open fields in the East Yorkshire parish of Butterwick in 1563, showing the remarkably regular arrangement of the furlongs and strips within them (M. Harvey)*

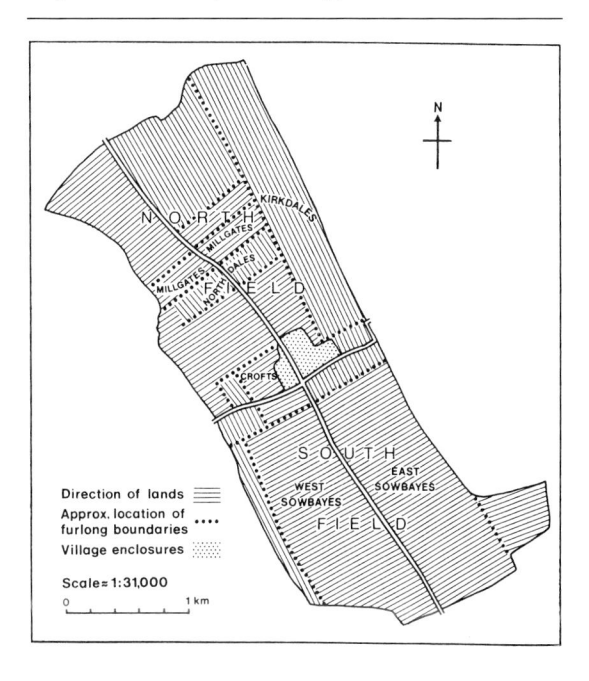

LONGHAM c.1595

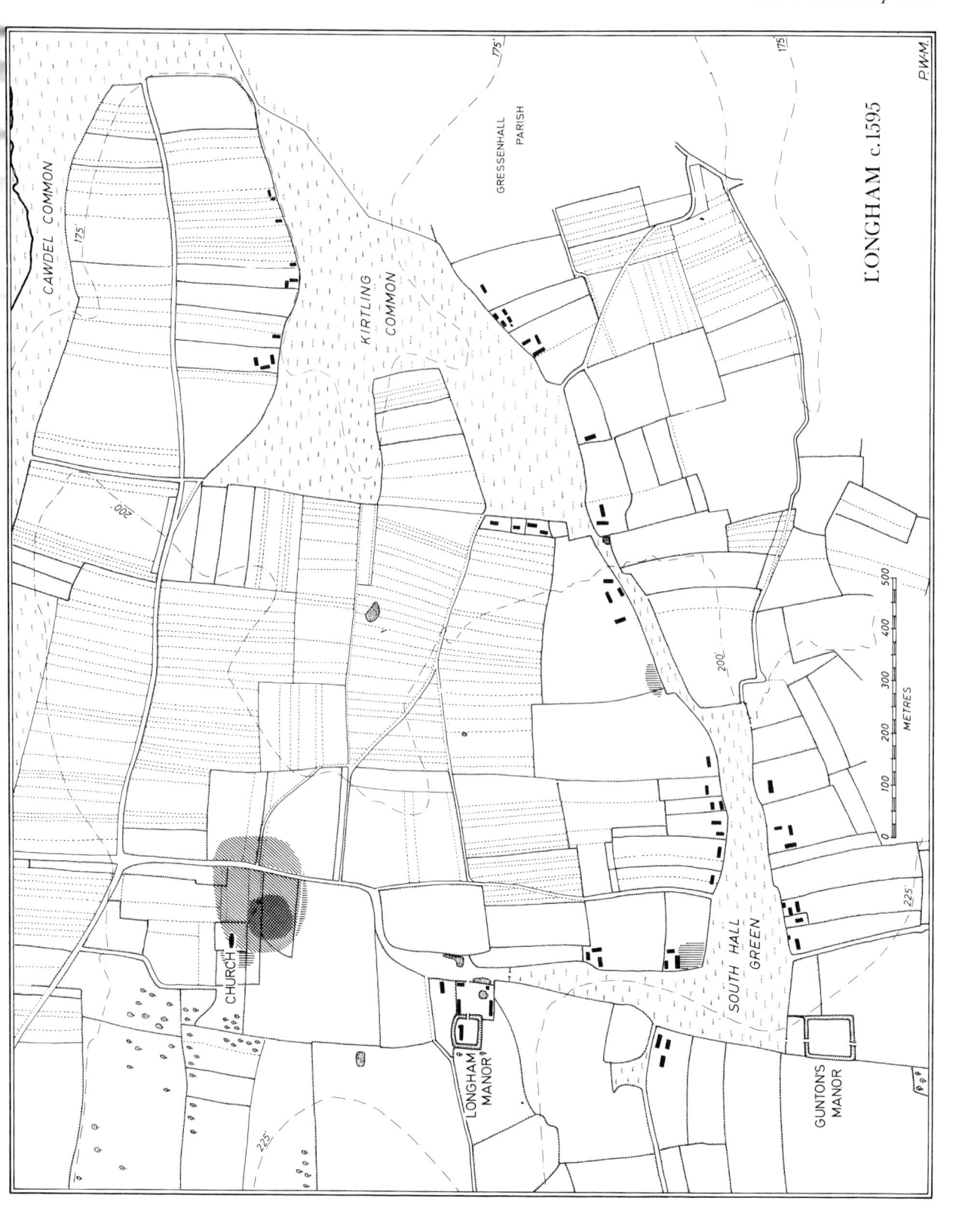

14 *Plan of Longham, Norfolk, showing the expansion of population between the early (double shading) and late (single sloped shading) Saxon period and then the shift of population to the village greens at the beginning of the Middle Ages (vertical shading). This distribution, based on pottery found in field-walking, has been mapped on to a sixteenth-century map of Longham showing the strips within furlongs and the house sites of that time. In common with most other villages, there is no sign of pre-Christian settlement on this site (Field Archaeology Division, Norfolk Museum Service)*

Only about fifteen per cent of the land surveyed in the Domesday Book was covered by woodland, and evidence from Anglo-Saxon charters and place names suggests that there had been little clearance over the previous three centuries. Most of England therefore was farmland with islands of wood, rather than the other way round. Areas of woodland did still survive in the Weald, on the Chiltern plateau, in north Warwickshire, east Cheshire, east Derbyshire and south-east Staffordshire, but even here the woodland was full of villages and hamlets, all practising subsistence agriculture.

All woodland was intensively farmed, either being managed to produce wood for building and fencing, or for pannage for pigs. This involved allowing the pigs into the woods in the autumn to forage for acorns for a few months before the pre-Christmas slaughter.

'The Anglo-Saxons in 600 years probably increased the area of farmland, managed the woodland more intensively and made many minor alterations. But they did not radically reorganise the wooded landscape'.[13]

Summary

There is little evidence for early Anglo-Saxon farming in the landscape. This was a period of population decline and therefore no radical reorganization. The reconstructed village at West Stow reminds us of the impermanent building materials and type of construction, which was typical of many early Saxon sites.

Probably by 1100 most village sites and parish boundaries, often marked by boundary ditches, had been established and so modern settlement patterns are based on the late Saxon agricultural unit necessary for self-sufficiency. Some isolated farms, too, were established by the time of the Domesday Book.

It is clear therefore that the foundations of the medieval farming landscape had been laid by 1086 with nucleated villages in the lowlands and many isolated farms in the uplands.

3 Early medieval agriculture and its field systems

Two main systems of farming existed in medieval England; that based on nucleated villages and open fields, and that on scattered farmsteads and enclosed fields. Scattered farmsteads with their own land, were more likely to survive in thinly populated areas, such as upland margins, and on heavy soils. However, we know far less about them than the better documented communal systems of the more densely populated midlands.

Nucleated villages are a prerequisite of communal farming within an open field system, and it is likely that it was a need for a community effort to intensify production as populations rose, which drove people together, either on their own initiative or as a result of organization from above. There is likely to be more documentation when this process of reorganization was initiated by the landlord, and written evidence tells us that both the church and lay owners were instrumental in relaying out their lands and creating planned villages in the twelfth and early thirteenth centuries. The Cambridgeshire village of Spaldwick was given a neat green and rectangular plan soon after 1209 when it was acquired by the bishop of Lincoln, while the bishops of Durham were responsible for the wholesale replanning of many of the villages they held in north-east England. Cerne Abbas in Dorset was probably replanned by the adjacent Benedictine abbey in the twelfth century. Lay owners, such as the Bardolf family in Lincolnshire, created Castle Carlton in the Lindsey marshlands and Riseholme in West Lindsey in about 1166.[1] Previously it has been thought that new planned villages were confined to the areas affected by William the Conqueror's harrying of the north. But it is becoming clear that planned villages, like planned towns, are to be found on the estates of enterprising land owners all across England, and with the replanning of the villages went the reorganization of the fields into some sort of open communal system.

However, this view of the overbearing hand of the lord must be tempered by examples which show the power and influence of the medieval community. A document survives describing the reallocation of land at Segenhoe, Bedfordshire, in the 1160s, as a result of unjust seizure of land during the anarchy of Stephen's reign. The document shows that this was carried out under the supervision of the manor court and 'six old men' rather than the authoritarian control of the lord.[2]

Many landlords were remote from their estates and visited them only intermittently. Most of the officials of the manorial court were tenants, and positions such as that of reeve and rent collector were elected. Their service, once elected, was often a compulsory condition of customary tenure. The power of the tenants was likely to be greater in communities controlled by many lords and Domesday book indicates that there were many such parishes.

15 *The village of Milburn, Cumbria planned around its green and surrounded by its strips, irregularly enclosed during the later Middle Ages (University of Cambridge Air Photograph Collection)*

16 *Planned village within its regular pattern of fields at Appleton-le-Moors, Yorkshire (University of Cambridge Air Photograph Collection)*

The regulation of the fields was the most important function of the village's internal governing machinery, and how this operated depended on many things, such as the way the land was distributed amongst families, the competition for land use and the amount of village grazing. There would, for instance, be a need for the careful regulation of grazing where pasture was scarce. Beyond this was the ever-pressing problem of an increasing population. It was these different social, demographic and agricultural pressures on the land that were responsible for the variety of types of field arrangements that developed.

The 'open-field' system

Much of our written and cartographic evidence for the open-field system dates from the final years of its operation, after the end of the Middle Ages, whilst the field evidence, in the form of ridge and furrow, also shows the system as it was finally practised, before being put down to pasture. Thus, we know far more about the end than the beginning of the system. It is not difficult, however, to show from a study of maps, manorial records and field names, that surviving examples of ridge and furrow and their named furlongs date back sometimes as far as the twelfth century. When this early evidence is combined with the field evidence visible today it is clear that frequently the overall physical layout of open fields changed very little until their enclosure.

The open fields, as developed by the medieval period, involved, first, the dividing of arable and meadow amongst the cultivators. Secondly, both the arable and meadow were thrown open at certain times for common pasturing. Thirdly, there was also waste land over which commoners exercised rights of exploitation, including grazing. All these activities were regulated by some form of village or manorial assembly.

In many open-field parishes there was very little common pasture and so the use of fallow for animal grazing was crucial to the system. If this were to be possible the fallow had to consist of a block of strips, so the typical three-field 'midland system', where one field was fallow and the two other were cropped, was likely to develop. Where pressure on grazing was less, then more flexible systems could be implemented. The fields of Northamptonshire are typical of the 'midland system' and their careful study has shown how strictly organized they were. Most had a regular order of tenants holding strips, so that the same neighbours were always adjacent. A single holding was called a yardland or virgate and consisted of a single ridge in each furlong. A two or three yardland holding therefore consisted of two or three ridges in each furlong, and from the thirteenth to the eighteenth centuries taxes and rates were assessed on the number of yardlands held. The highly ordered layout of field systems, some with long strips, such as in Holderness, which imply a high degree of central control when they were laid out, and the physical relationship between fields and early taxation systems all point to a single planned operation in the creation of each township.[3]

The open fields were often based on pre-existing systems of agriculture and field shapes; they did not evolve in an empty countryside. For instance, even the field boundaries of the remarkably regular strip fields layed out at Wharram Percy and the adjoining parishes respect the layout of the Romano-British estate boundaries that preceeded them.[4]

The open fields were divided into furlongs and the crop rotations practised were based on the individual furlongs rather than the entire open field. Wheat, barley, oats, rye, beans and flax were all grown, often in complex rotations, but when the land was left fallow, it was the whole field, rather than the individual furlongs, that was

Early medieval agriculture and its field systems

left and the village livestock was free to graze across it.

The working of this system had probably reached its most ordered by the twelfth century. However, variations would quickly develop, partly as a result of inheritance and exchanges of land. In areas where land was shared equally amongst all the sons, it would be far less stable than where only the eldest child inherited. As these customs varied from area to area and from one class of society to another, it is not surprising that a wide variety of types of form and operation of field systems were being operated by the fourteenth century.

The most common field evidence for these open-field systems is the survival of ridge and furrow, which in spite of modern 'prairie' arable farming, is the most ubiquitous of all archaeological remains of the farming landscape. These low parallel ridges between ditches run down the steepest gradient under the hedges of later enclosure, to the limits of each medieval furlong within the open field. They are formed by the continuous turning of the sod towards the centre and away from the ditch by a heavy fixed mould-board plough. By the eleventh century, ploughs with coulters and mould-boards as well as a share were

17 *Ridge and furrow surviving within Wimpole Park, Cambridgeshire, in an area that has not been cultivated since the eighteenth century (University of Cambridge Air Photograph Collection)*

18 *Aerial photograph showing the open fields of Laxton, Nottinghamshire. The reversed S of many of the strips can clearly be seen (Crown copyright MOD)*

used over most of England. They could break up the ground more thoroughly than earlier light versions but had to be pulled by six or eight oxen yoked in pairs. The turning room needed for this cumbersome implement meant that a curve, or reversed S tended to develop along the ridge, and so much of the midlands of England were, at least until recently, covered with these curving waves of ridge and furrow. Sometimes they have become fossilized in modern field boundaries. At the end of each strip the plough was lifted out to turn it, dropping a small quantity of soil as it did so. This gradually built up into a raised headland and the survival of these headlands allows for the identification of the different furlongs.[5]

Each ridge does not necessarily represent an individual strip, which could have been made up of several ridges. Ridge and furrow did not exist in every area where we know there were open-field systems. For instance, in much of Norfolk and Suffolk where sixteenth century map evidence shows that the land was farmed in strips, and where much of the heavy clay land was put to pasture at the end of the Middle Ages, little evidence for ridge and furrow remains. We must assume that Norfolk and Suffolk techniques were deliberately designed to keep a flat surface. The creation of ridge and furrow simply does not seem to have been part of the traditions of cultivation away from the midland heartland of the practice.

This may seem surprising, particularly as much of East Anglia's ill-drained boulder clay might be expected to have benefitted from the drainage provided by the furrows of ridge and furrow which, it is generally accepted, was an important drainage technique before land drains. In fact, it continued to be used as such until the nineteenth century in some areas. These late examples can be distinguished from medieval ones because the lighter nineteenth century ploughs did not need to follow the medieval method of using a large area to turn, and so are straight rather than following a reversed S curve.

In the midland region the strips averaged about a third of an acre (7×180 metres), although there was considerable variation in size. For instance, in some areas such as the

Yorkshire Wolds, strips could be as much as ⅝ mile (1 kilometre) in length.

The study of the distribution of surviving ridge and furrow, rather than the open fields shown on maps, usually not dating from before the sixteenth century, is important. It is evidence for the extent of medieval agriculture at its greatest, during the period of population pressure, reaching a peak at the beginning of the fourteenth century. Large areas of slight ridge and furrow survive on the chalk downlands of southern England where cultivation did not go on long enough to obliterate the Roman and prehistoric fields underlying them. 'Without the recognition of this ridge and furrow, which is not depicted on any maps, an important part of the medieval agricultural history of the area would have been lost'.[6]

Whether ridge and furrow was typical of late Saxon farming is unclear. Some examples, fossilized under Norman remains, have been found at Hen Domen in Wales and Gwithian in Cornwall. At Bentley Grange, South Yorkshire, mining spoil heaps of the twelfth century overlie ridge and furrow. At Titchmarsh in Nottinghamshire, a manorial garden was enlarged to include part of a furlong, but all these are no more than isolated examples.[7]

A modification of the strip system found in hilly country is the strip lynchet, apparent as long parallel terraces along steep hillsides, often of considerable size and complexity. They were a way of cultivating steep land in a time of land shortage. They vary in length from 77 to 220 yards (70 to 200 metres) and differ from the earlier Roman ones in that they do not have a squared end, but rather fade out in the area where the plough turned round. They are very difficult to date and some were still used in the nineteenth century. In upland contexts, such as Upper Wharfedale, Yorkshire, where the hillsides have been sliced again and again by lynchets, pressure for land resulting in the cultivation of such remote upland areas is only likely to have occurred in the early Middle Ages and so they are probably pre-fourteenth century in date.

It is unlikely that the open-field system, with all the communal custom that went with it, was fully established until 1200. Certainly the replanning and relocating of villages, which played a fundamental part of farming change, continued through early medieval times.

Whilst individual open fields survive on the Isle of Axholme (Kent) and at Braunton, Devon, where in 1950 twenty-two proprietors held strips of between ½ and 2 acres (under a hectare) in a 350-acre (140-hectare) field, only at Laxton in Nottinghamshire does an open-field parish, with all the communal institutions that the organisation of its farming involved, remain, almost in its entirety. Its 540 acres (218 hectares) of open field are still administered by a field jury responsible to the court leet, consisting of all the farmers. The village was mapped in 1625, when the fields were at their greatest extent. 1,300 acres (530 hectares) of ploughland were in open fields, 1,500 acres (600 hectares) in enclosed closes, some of which were ploughed, 240 acres (100 hectares) were meadow and 200 acres (80 hectares) in common. There are eighty-seven dwellings marked on the map. This was as large as was practical for an open-field village to be. Outlying strips could be as much as 3 miles (5 kilometres) from the village street where all the farmsteads were. A late nineteenth century comment must have been as true in the Middle Ages: 'A tenant holding a farm of about 200 acres would require the greater part of a day to inspect it.' The time was coming when farms at the margins would have to be put in separate farms. Although this process began in the eighteenth century one of the most striking features of Laxton today is the number of farms concentrated in the village itself, rather than scattered around the fields.[8]

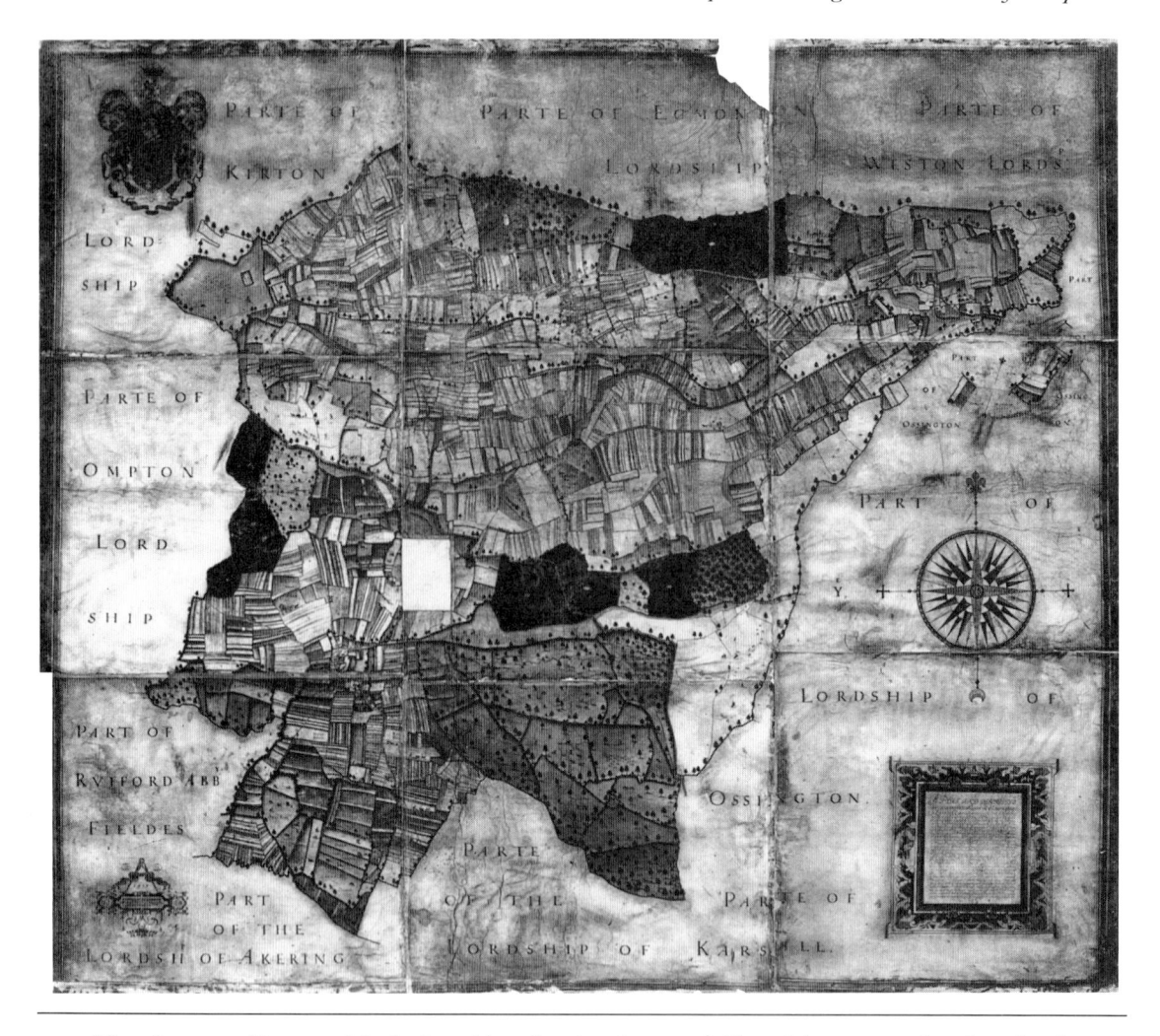

19 *The 1625 map of Laxton, Nottinghamshire, showing the open fields and the strips within them (Bodleian Library Oxford)*

Other medieval field systems

The classic open-field system was not found all over Britain. It was most typical in a band through the lowland centre of England, through into eastern lowland Scotland, as far north as the Firth of Forth. Norman penetration of the lowlands of south and north Wales and the Marches brought open fields to parts of Wales, but they are only rarely found in Devon or Cornwall.

In Essex there were few open fields. The present-day landscape is the result of the gradual simplification of the medieval landscape by the removal of hedges, woods, roads and nearly all the heathland. It has few deserted villages because nucleated villages are normally associated with open-field agriculture and so did not exist. Instead there are many deserted farmsteads and moats.

The most usual non-open system was the infield-outfield which was more suited to an area concentrating on animals rather than

Early medieval agriculture and its field systems

cereals. Because of the lack of documents it is easy to overlook its importance over much of Britain. Here small enclosed infields would be cultivated individually every year. They would not be left fallow, but would be well manured with dung and household waste. The much larger outfield would only be cultivated occasionally and was otherwise used for pasturing animals. It is likely that this is the most ancient system of cultivation and that many areas that later grew to become open-field villages were originally infield-outfield ones. As population pressure grew, and more cereals were needed, they adopted the newer, more intensive system.

Physical limitations resulted in the survival of the infield-outfield system on the poor lands of the Norfolk and Suffolk brecks and the uplands of Somerset, Devon, Yorkshire, Northumberland, Durham,

20 *Aerial photograph showing the Laxton farms concentrated along the village street. Behind the farms can be seen earthworks and then the individual enclosed pastures with ridge and furrow, showing that they were previously cultivated. Beyond and out of the photograph are the open fields (University of Cambridge Air Photograph Collection)*

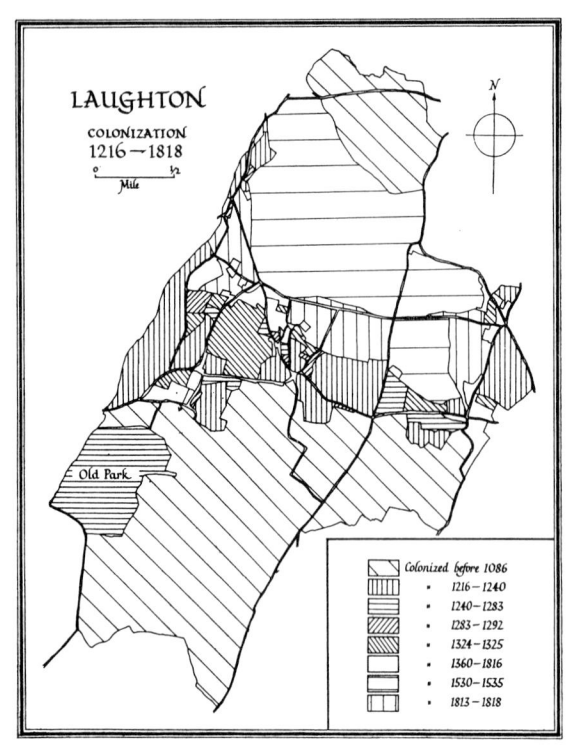

21 *Map of the parish of Laughton, Kent, showing the gradual assarting of new land throughout the Middle Ages (J.S. Moore)*

much of Wales and almost all of Scotland.

In areas such as much of Cornwall, where an infield-outfield system operated over most of the area in the thirteenth century, the small field is typical. Here there are probably more hedges per square mile than in any other county. Many of the small fields created were of prehistoric origin, but the traces of an infield-outfield system can be detected. Amongst the maze of field boundaries, the wide sweeping curves of the original outfield walls, with the later enclosures inside them, can be traced, both on the ground and on maps.

In the Weald of Kent and in Surrey where dispersed settlement and partible inheritance was usual, a mixture of enclosed fields and small blocks of strip fields, or 'yokelands', which were gradually fragmented but which were not always farmed in common, developed.

In the Wealden parish of Laughton the open fields are thought to be early (pre-eleventh century). In contrast to some other areas studied, Laughton's open fields did not increase in size after 1020. As new land was brought into cultivation it was worked in compact blocks by individual farmers. Gradually the open-field system was abandoned, in the thirteenth century. In other areas, such as the woodlands of the midlands, where new land was being brought into cultivation through the early Middle Ages, a mixture of irregular open fields and later enclosures was typical.[9]

Evidence for field systems and boundaries is often fossilized in the modern landscape, sometimes in unexpected places. It is possible to get a rough idea of the age of a hedge by noting the species of trees in it. The older the hedge the more species it is likely to contain. In 30 yards (27 metres) of hedge each species represents about 100 years. Devon and Kent, for instance, were well hedged by the late Middle Ages and so five species and upwards are now found in many of their hedges. They also have a significant number of three- and four-species hedges, the result of post-medieval expansion on to moor and heath, and a small number with two or three species, the result of later enclosure. One of the most varied hedges recorded is in Felsham, Suffolk, a parish of ancient scattered fields and a Roman or Iron-Age rectilinear field system where three hedges have thirteen species. No doubt more await discovery. This method of hedge-dating allows for the distinguishing of hedges planted during the period of the enclosure acts (*c.* 1770–1820), those established in the Tudor or Stuart periods, and those of the Middle Ages, but nothing more precise. The system also breaks down for hedges over 1,100 years old. It is impossible therefore to distinguish

Roman or Iron-Age hedges.[10] Research such as this can throw up some surprises, such as, for instance, which modern property boundaries in housing estates are also those of older fields.

Livestock

Both open and closed field-systems relied on livestock to keep up their fertility. Whilst the crops were growing livestock would be kept on the common or outfield, but would also be grazed on fallow land or stubble. Animal-bone evidence suggests that sheep were the most numerous of domestic animals, not only on the monastic and other estates famous for their wool production, but also on the peasant farms. Young stock found its way into the town butchers' shops, while older animals, kept for breeding and their wool, finally provided the rural population with its meat. Milk and cheese were also important by-products.

Cattle too were not only kept for their meat, but were a valuable draught animal and it is mature bones, often showing signs of a hard working life, which are found in excavations. An increase in the number of younger animal bones found for the latter Middle Ages suggests that they were being increasingly bred for meat.

Pigs are the third domestic animal whose bones form a significant part of medieval archaeological deposits. They were useful browsers, eating household left-overs and often fattened in the woods on acorns, in the autumn. They seem to have declined in importance, perhaps as the area of woodland declined during the later Middle Ages, and lamb, and especially mutton was providing an increased proportion of the diet.

These domestic animals were mostly smaller than their Roman counterparts, which were in turn much smaller than modern animals. This could have been the result of the shortage of grazing and overstocking early in the period, but if this was the case one might expect an increase after the Black Death. However this did not happen.[11]

It is against this background of a mixture of arable and pastural farming, operating under a great variety of systems across the British Isles, that we can see the general agricultural progress of the Middle Ages.

Summary

The remains of medieval field systems can often be seen in the landscape. Where land has been pasture since the Middle Ages, ridge and furrow may still remain. This is most likely to survive on the heavy clay meadows of the midlands, in areas of upland not cultivated since the Middle Ages, and in land included in parkland, so escaping ploughing ever since. In the anciently enclosed areas of the country, species-rich hedges are an indication of their age, and an afternoon spent species-counting will soon reveal which are the medieval and which are the later boundaries.

4 The period of medieval expansion

The Norman Conquest

It is clear that the England taken over by William after 1066 may have contained a farming system in transition. The population was increasing and by Domesday (1086) was probably about two and a half million. This extra pressure on land, plus the fact that some areas were suffering from soil exhaustion, having been cultivated continuously for several thousand years, meant that more intensive methods were needed in the more populated areas. Here a change to nucleated villages and communal farming was being made, whilst elsewhere, isolated farms remained, and new single farmsteads were still being set up by colonists, particularly in the wooded areas of Arden, the Weald and later the New Forest. This simple explanation of the different regional distribution of nucleated settlements and isolated farms does not tell the whole story. Areas of the south-east with soils similar to those of the midlands, and which were heavily populated from the early Middle Ages, have retained a dispersed pattern of settlement in contrast to the classical 'midland system'. For an explanation of this we may have to look back to the early post-Roman period and the establishment of the first Saxon kingdoms.

The document that has long been seen as the basis for further study of the agriculture of post-Conquest England is the Domesday Book. Compiled in 1086, it was not intended as a historical document. It was a survey of resources, indicating the value of land and its produce for tax purposes. Norman scribes and commissioners imposed their own preconceptions and terminology on the realties of Saxon and Scandinavian England. The methods by which these resources were being exploited, or the organization of agriculture, are not revealed. It has been argued, for instance, that 'bordars' as described in Domesday, were a freer class of people who had taken in land from the waste outside the communal farming system, and who lived an independent life, in isolation, on their smallholdings at the edges of commons, wastes and woodlands. As such, it is suggested, they were a symptom of agricultural expansion.

Domesday does, however, show us an evolving landscape in which any under-used areas were fast being exploited. The area under plough was as much as eighty per cent of the 1914 figure. Thirty-five per cent of England was devoted to arable, twenty-five per cent to pasture and fifteen per cent to woodland.

It is clear from Domesday that it would be wrong to associate all isolated farms, particularly those in the uplands of Britain, with colonization of new land in the twelfth and thirteenth centuries, as the population grew. For instance, the Domesday entry for the tiny, steep, coastal parish of Culbone on the edge of Exmoor lists land for two ploughs, 100 acres (40 hectares) of woodland and 50 acres (20 hectares) of pasture. The inhabitants included two 'villeins', one 'bordar' and one 'serf'. The parish today consists of a church and two isolated farms, which can perhaps be identified with those of the two 'villeins' in Domesday. In the much larger neighbouring parish of Porlock, six 'villeins' are recorded. Five of the nine isolated farms in the parish can be traced to the early Middle Ages through such manorial documents as bailiffs' accounts, Pleas of the Forest and Subsidy Rolls, and it is probable that the majority of the nine date from before 1087, as farms worked by the six villeins.[1]

Although some of the open-field parishes with their nucleated villages and ridge and furrow were already in existence by 1066, many more were laid out in the early Middle Ages. Where the fields already existed, efforts were often made by the community to bring more land into the system. We can see the results of this continuing colonization both in the landscape and in early maps. What we cannot see is the process by which it took place. How far was it the result of a community effort, and how far was it imposed by a feudal landowner? The fact that many parishes were divided into more than one manor whilst having a single field system, suggests that the farming was community, rather than landowner, controlled. The example from Segenhoe, Bedfordshire, already quoted suggests that the manor court representing the farmers of the open field was indeed still very influential.

Manor court records tell us a lot about the types of farming practised in the open field and they have been widely used by Christopher Dyer and Bruce Campbell in their reconstructions of medieval farming. Wheat was the preferred crop, amounting to half the food grown on eighty-three estates studied by Dyer.[2] It was also the main charred-food cereal found in medieval pits in Winchester. Barley was grown as a spring-sown crop and increased in importance after 1325. It was either used for malting or a more primitive type called bere used as a coarse bread flour in Scotland, north-west England and parts of Wales and Cornwall. Oats were most important in the north and west where it was a food grain as well as an animal fodder. Rye was widespread, but the least important grain, being particularly suited to light soils. Peas, beans, lentils and vetches were all grown, and increased in quantity in the fourteenth century. Hemp and flax were also of local significance.

Cattle and sheep were the most important animals kept, although there were some huge pig herds of over 100 kept by monasteries. In spite of the great monastic flocks of sheep, more were probably kept by peasants. Between 1280 and 1290, the wool of three million sheep was shipped from Boston, Lincolnshire, and two million from London.[3]

Bread wheat was the most widely cultivated cereal. Rivet wheat was popular during the early Middle Ages. It needs a warm climate to be cultivated successfully, and so perhaps is an indicator of the fact that this period of population expansion was warmer than at present, providing better conditions for farming than have prevailed since.

Part of every village system was the 'demesne', or home farm, of the lord of the manor, who might be resident, but more often was not, and so it would only be visited occasionally by the owner. The Earl of Norfolk's estates at Forncett, south of Norwich, were run by a bailiff who lived in the manor house. Besides this sizeable residence there were outbuildings containing a dairy, three stables, a cattle house, granary, hay-house, goose-house, hen-house, and pin-fold, built of clay with straw roofs.[4] The bailiff was frequently visited by a travelling steward who presided over the manor court.

The bailiff was required to keep detailed accounts; many of these survive although few have been studied. The demesne land of the estate could well be scattered across the manor and intermingled with that of the tenants. At Forncett the demesne totalled about 300 acres (120 hectares). As well as a single close, Hall Close, near the village, there were small assarts, or newly enclosed land, on the edge of the village and scattered acre and half-acre strips. If some of the lord's land was in the open fields he would be as much at the mercy of local traditions as anyone else. In the early Middle Ages, therefore, he had little influence over the farming or landscape of a community governed by custom.

Wales and Scotland

The situation was slightly different in Wales and Scotland. Much of south Wales was covered with an anglicized open-field system and nucleated villages surrounded by ridge and furrow. By the fourteenth and fifteenth century demesne land was being let and bondsmen were transformed into rent-paying tenants who began to break up the open fields into independent farmsteads. In contrast, on many of the hillsides, Welsh freemen lived in scattered farms. To the north, the hamlet was the typical form of settlement with communal farming based on shared-plough teams. The ideal group, according to a fifteenth-century document from north Wales, consisted of nine houses with one plough, one kiln, one churn, one cat, one cock and one herdsman. In the Middle Ages cattle were more important than sheep in the Welsh farming economy, and the uplands were supporting large numbers. The 650 acres (260 hectares) of moor at Hafod Elwy, Denbighshire, supported 180 in 1334.[5]

The situation was complicated in Wales by the custom of partible inheritance where both freehold and leasehold land was subdivided amongst all the sons. This fragmenting of holdings could reduce a well-established family to penury over the generations, and finally result in the disintegration of the landholding pattern. It also led to the need to set up new settlements and cultivate the waste. An act of parliament of 1542 provided for the abolition of partible inheritance upon the next change of heirs. However, evidence for the tiny strips of land created in the Middle Ages survived and were still visible in the village of Llanynys in Denbighshire until they were ploughed out in the 1970s. As late as 1840 sharelands lying in strips remained to be marked on the tithe map of Pennant in Merionethshire, some of which lay at heights over 800 feet (245 metres).[6]

Very little is known about the system of landholding in Scotland before the eighteenth century, but here too farming appears to have been based on a small hamlet or 'township' worked communally by four to six tenants. The strips were located in an infield and outfield, about a third in the infield and two thirds in the outfield. The infield would be permanently cultivated. It was never left fallow and all the manure from the settlement was put on it. Crops of barley and oats were grown in rotation. The outfield formed irregular patches further from the townships. Here oats would be planted after animals had been folded on the land overnight. A crop would be grown continuously for about three or four years until the fertility dropped; then the outfield would be left to rest, probably for about five years. Famine was frequent and the population grew only slowly, so that this primitive form of agriculture prevented excessive pressure on the land. Most of the field evidence for this type of farming was obliterated by later enclosure, but the head dykes which enclosed the fields, forming a boundary bank, can sometimes be traced.[7]

The period of medieval expansion

The influence of the Church

By far the most powerful landowner in medieval Britain was the Church, which by the early fourteenth century owned about twenty-five per cent of the land. Only the Church had the resources to tackle major land improvements such as drainage. The priory of Canterbury started the improvement of the pastures on Romney Marsh, while the monasteries at Ramsey, Ely, Spalding, Crowland and Thorney all contributed to work around the Wash, particularly in the peat fens where land was reclaimed for pasture and then let for a money rent rather than labour services. The records of Ramsey Abbey show that in the year 1206–07 twenty-eight tenants were working newly reclaimed land in Upwell and Outwell.[8]

Much of the Somerset Levels was also owned by the Church and the abbeys of Muchelney and Glastonbury were responsible for reclamation for pasture along the Parrett valley. Here, low-lying meadows were surrounded by walls which could then be used as causeways.

Woodland clearance was also undertaken by the monastic estates. 1,000 acres (400 hectares) owned by the Bishop of Winchester at Witney, Oxfordshire, was cleared in the first half of the thirteenth century, but clearance on this scale was unusual and it was normally small areas that were brought into cultivation.

The most important monastic order concerned with reclamation was the Cistercians who made great efforts to consolidate their possessions and manage them directly rather than rely on rents from scattered holdings. The order was founded in 1098 and by the Reformation there were 100 Cistercian houses in the British Isles, two thirds of which were founded by 1152. They preferred remote sites, extending the existing pattern of monastic settlement to the north and west. The founding of an abbey included extensive programmes of clearance and enclosure of waste as well as the establishment of many of the farms on the hills and flood plains that have survived intact as units to this day. The Cistercians particularly favoured areas that had been described as 'waste' or 'largely waste' in 1086, and exploited these lands by setting up 'granges', or isolated farmsteads, worked either by monks or lay brothers under the direct control of the monastery. Nearly half of known twelfth-century granges were on newly reclaimed waste. Huge flocks of sheep were also kept on many of the estates. Meaux Abbey owned 14,000 sheep in the thirteenth century which it kept on its Yorkshire estates.[9] The Fountains Abbey Grange at Malham was surrounded by small meadows for hay and enclosed pasture for sheep. There were also two large sheep houses, one of which has been excavated. This was a rectangular building beside a yard with an open lean-to. Sheep housing was widespread in medieval Europe and continued in use in the Lake District until

22 *Reconstruction of Evesham Abbey Grange, Gloucestershire (C.J. Bond)*

the eighteenth century.[10]

The typical grange was predominantly an arable or mixed holding of between 300 and 1,000 acres (120 and 400 hectares). Cattle were as important as sheep in the early Middle Ages. Cattle lodges or 'vaccaries' were found on monastic estates across northern England in the Lake District, Northumberland, Yorkshire and Lancashire, but nowhere further south. Many granges were intensive dairy farms and at Gatesgarth, at the head of Buttermere, there was a cowshed 67 feet (20 metres) long for housing forty cows and their young in the winter as well as enclosed high fell grazing and meadows on the valley bottom. Fountains Abbey granges included vaccaries where cattle were moved from estate to estate and hides exported through Newcastle.[11]

In the Lake District, peasant holdings were confined to the valley floors where the only possible arable land was to be found, whilst landowners grazed the uplands; of these landowners, the Cistercians were the most important, based mainly on their abbeys at Furness and Holme Cultram. Sheep and their wool were by far the most important product, with one monk at Holme Cultram responsible for the sale of the monastic wool as well as acting as middleman for local farmers. Small farms in Borrowdale, many of which still exist, were tenanted and their business co-ordinated by local monastic granges. The hamlet of Grange-in-Borrowdale was founded by Furness and a grange at Hawkshead oversaw the land between Coniston Lake and Windermere.[12]

The farmsteads provided shelter for both humans and livestock and stood at the centre of consolidated holdings. Some of the wealthier monasteries controlled many granges. There were seventy-five in Yorkshire by 1200, of which twenty-four were owned by Fountains Abbey, but this was exceptionally large. Remains of these farms above ground are limited and very few of the farms now called 'Grange Farm' have anything to do with the Cistercians. Excavations at the Knights Templars site at South Witham in Lincolnshire, built between 1164 and 1185, revealed three large-aisled barns in the main outer court as well as three smaller buildings for animals, and kilns for corn drying and brewing.

Whilst the period of large-scale direct exploitation of their granges by the Cistercians probably did not last more than 200 years before they were let as tenant farms, many of the wealthy Benedictine monasteries ran their estates in more or less a consistent way for up to 600 years. For instance, from the eleventh century Abingdon Abbey had extensive estates in Berkshire and Oxfordshire. Orchards, gardens and vineyards are recorded near the abbey, whilst ridge and furrow testifies to arable farming, with rich meadows and pastures along the Thames. There are references to enclosed pastures after 1350, and to hedges being planted at Culham in 1355–6. Further afield, in east Berkshire, the abbey owned woodland which it exploited. The estate covered about twelve manor houses with barns and dovecots. Mills, fish ponds, and quarries are also mentioned.[13]

On the Evesham Abbey estates, fourteen barns, twelve dovecots and eighteen thirteenth-century fishponds survive. From the documents, it is clear that these estate farms were very substantial, containing large, stone-built porched barns, dovecots, granaries and sheds for livestock. Not surprisingly, it is on these monastic estates that some of our oldest and finest farm buildings remain.[14]

At Cressing Temple, Essex, are two of the earliest timber-framed aisled barns in Britain. In 1137 the manor of Cressing was granted to the Crusading Brothers of the

23 *The interior of the thirteenth-century aisled barley barn at Cressing Temple, Essex (RCHME)*

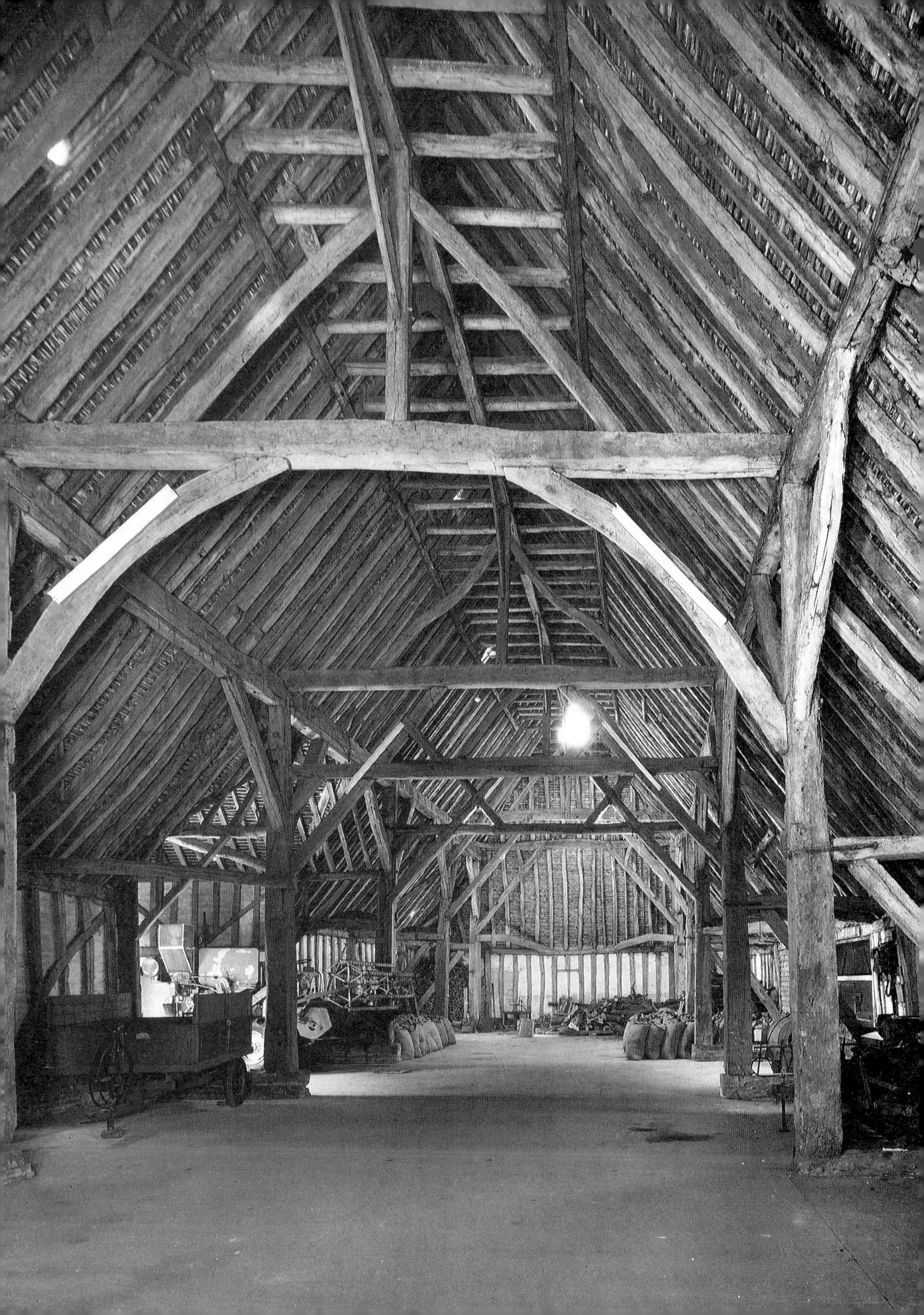

Knighthood of Solomon's Temple at Jerusalem (the Knights Templar), and later they were granted land in neighbouring Witham. By 1300 they had estates in the area of about 2,000 acres (800 hectares). Cressing therefore became the centre of a major agricultural and commercial enterprise. Only the two great barns survive from what must have been a large complex of buildings. The earliest was built about 1200 and the other sixty years later. Both are aisled structures, the only way in which early medieval builders could span a wide space. These barns are 48 feet (14.6 metres) and 44 feet (13.4 metres) wide. Documentary evidence describes a dairy, brewhouse, cider mill, granary, dovecot, and smithy. Stock included cattle, pigs, geese, chicken and peafowl. Gradually the order fell into disrepute because of their wealth and extravagence, and as their *raison d'etre* ended with the Crusades they were finally dissolved after 1308.[15]

Another fine timber-framed aisled barn survives on a second monastic farm in Essex, this time on the Cistercian grange at Coggeshall which was founded by King Stephen in 1140. The barn is now owned by the National Trust.

Equally large stone barns survive on other monastic estates, most notably those of Beaulieu Abbey in the New Forest and in Oxfordshire. Great Coxwell barn, Oxfordshire, was built by the Cistercians of Beaulieu in about 1300. The stone-aisled building, described by William Morris as 'beautiful as a cathedral' housed the produce of the grange farm. 'The great lines of its simple mass, the intersecting bodies of its two large transepted porches, the steep ascent of its gables and the noble silhouette of its vast roof are unsurpassed by any structure of like design . . . In the interior, this barn displays one of the most magnificent frames of medieval timber ever known in a building of this construction

24a and b *Reconstruction of the interiors of the Cistercian barns of St Leonards, Hampshire (left), and Great Coxwell, Oxfordshire (Walter Horn and E. Born)*

type'.[16] The farmstead, of which this barn formed part, was a large-scale enterprise with a regular staff which in the fourteenth century included a swineherd, eight ploughmen, two carters, a cheese maker, baker and forester.

Of the barn at St Leonards, on the abbey's Beaulieu estates, only half is now in use, but originally, it had the largest storage capacity, 52,6590 cu.feet (14,913 cu.metres), of any English barn.[17] A third barn at Shilton, Oxfordshire, was also owned by the Beaulieu monks, but of their other twenty-four barns, nothing remains. They are an indication both of the power and wealth of the church and of the importance of agriculture, in this case the production of huge cereal crops, to their economy.

As well as these fine aisled barns there were others of cruck construction, often also on monastic estates. This strong form of construction consists of a series of pairs of carefully chosen curved timbers set up to form an inverted V. The apex of the V takes the ridge and the roof, which, with its purlins and rafters, is supported by the crucks rather than the walls of the building. A thirteen-bay cruck barn, built for Gloucester Abbey at Frocester between 1294 and 1306 survives, whilst four barns on the properties of the abbey still stand. All are of cruck construction and were built, rather surprisingly, in the late fourteenth century; a time when farming was in decline and great estates were reducing the area of their demesne. The abbey barn at Glastonbury houses the county's rural life museum. The largest surviving cruck barn is at Leigh Court, near Worcester, and was built for the monks of Pershore Abbey.[18]

Expansion of the cultivated land, 1100–1300

The fullest exploitation of land was exemplified on the monastic estates, but is only the most extreme example of a movement towards greater efficiency on existing farms and moves outward into marginal country, which form such an important theme of the agricultural history of the years 1100–1300. The population at least doubled and probably tripled, and had to be fed without recourse to imported food.

The greatest population growth took place in areas of greatest agricultural productivity, as exemplified in the area around the Wash. We have already seen the importance of the great abbeys in developing this area, but individuals and communities were also busy from the tenth century reaching a peak of activity in the twelfth. By 1300, 10 square miles (30 square kilometres) had been reclaimed on the seaward side of the silt belt and a further 100 square miles (300 square kilometres) of peat fen inland. Prior to the drainage of the silt fen the area was very much poorer than the surrounding upland. However, with the Lay Subsidy (a document compiled for tax purposes in 1334), it was much richer; a fact to which the medieval churches of the fen edge are testimony. The medieval landscape of Marshland (a group of parishes in the Norfolk fens to the east of Wisbech) has been reconstructed by R.J. Silvester and the expansion into the silt fen can clearly be seen. The original fields around the villages are irregular, but the laborious digging of a myriad of parallel ditches to create minute, narrow, rectangular fields in the fen can be seen. Droveways stretching back into the village were left so that cattle could be driven up to the Smeeth, an open inland area of marshland that survived until the nineteenth century.[19]

As well as expansion into fens there was also colonization of marginal moorland. The nibbling away of Holme Moor on Dartmoor began in the eleventh century, but the deforestation of much of the area in 1204, as a result of increased population pressure, allowed for more expansion. The

48 · The period of medieval expansion

MARSHLAND The Medieval Landscape

25 *The medieval landscape of the fenland parish of Marshland, Norfolk, showing the gradual colonization of the grazing area of the Smeeth and the drove roads providing access to it (Field Archaeology Division, Norfolk Museums Service)*

earlier enclosures were divided into strips and used for arable whilst the later ones were created with stock-proof boundaries and droveways, and so must have been for livestock. Unlike the earlier reaves which were obviously part of a large-scale layout, the pattern of these boundaries indicates a piece-by-piece encroachment onto the moor. The whole area was abandoned in late medieval times.[20]

The period of medieval expansion

The Lake District is one of the few areas, other than Dartmoor, where a systematic study of the farming landscape, as distinct from the settlements within it, has been undertaken. A landscape survey team, working for the National Trust, has recorded all the man-made features on Trust properties. Careful study of the details of dry-stone walls, such as blocked entrances, builds up a chronology of assarting (taking land in from waste) as well as revealing early farms and communal boundaries. The word, 'assart', is derived from Old French and only came into general use in the late twelfth and thirteenth centuries. The boundary between open fell and enclosed land was marked by a stock-proof fence or wall forming a division between the tenanted land of the valley bottom and the communal grazing. The area thus enclosed gradually increased during the twelfth and thirteenth centuries. About a third of this wall around Great Langdale still stands, in some places to its original height of 5 feet (1½ metres). It ran in a great arc around the head of the valley for about 4 miles (6½ kilometres). By the thirteenth century there were fifteen tenants, each with about 9 acres (3½ hectares), their farms scattered along the valley bottom. In the enclosed area the

---

26 *Map of part of Langdale, Cumbria, showing the three medieval farms, the in-bye land with its irregular fields and surrounding ring-garth, and the large Napoleonic enclosure fields, set on steeply sloping land in the open pasture. The sheep folds are mainly along the ring garth (based on work done by a National Trust landscape research team)*

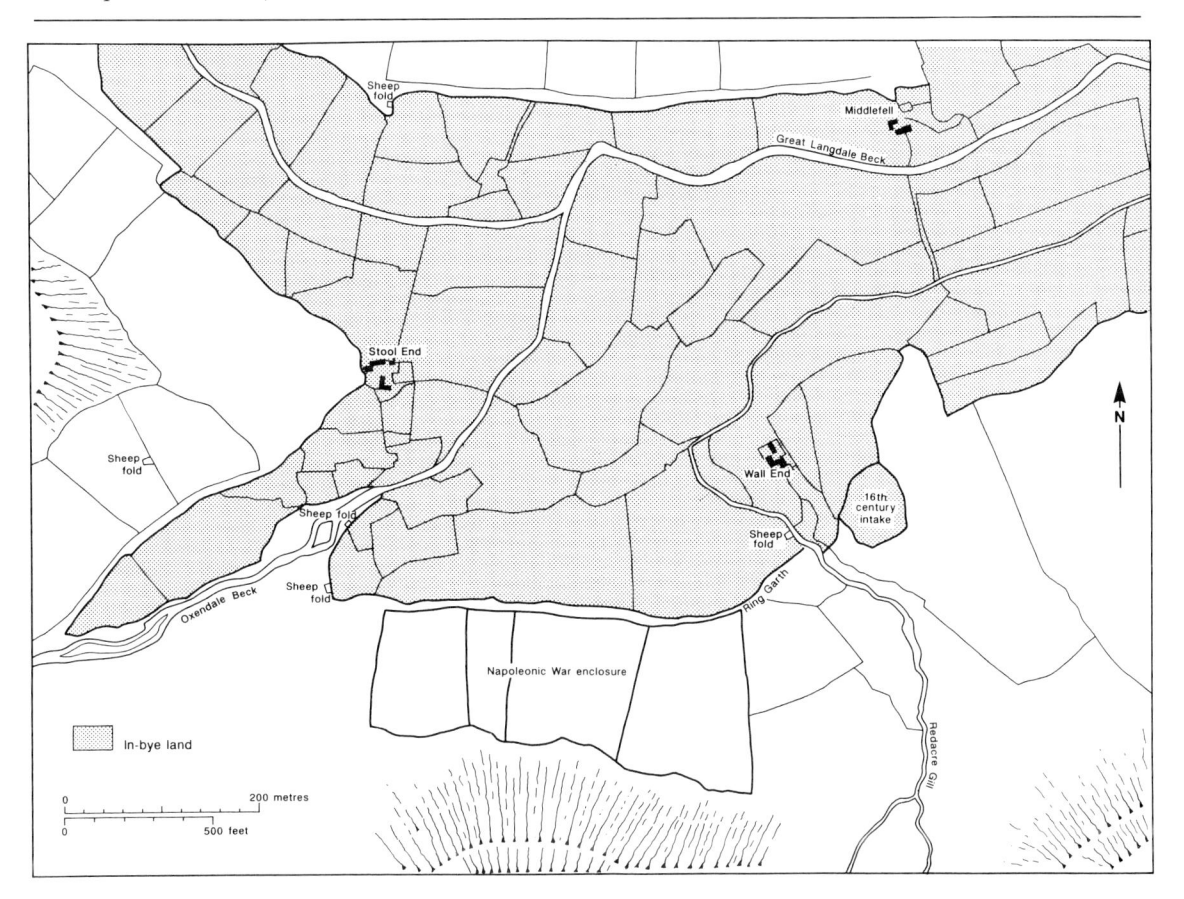

---

land was cultivated on a communal system with strip fields which were opened for stock in the winter. Most of these common fields were subdivided by the seventeenth century.[21]

In every area where there was forest there are records of clearance. Much of this was land where woodland had regenerated since first being cleared in the Iron Age or Romano-British times. Thiston and Rockingham Forest, north Hampshire, and Wychwood, Oxfordshire, as well as Pendock, south-west Worcestershire, are all areas where there was a proliferation of small manor houses and moats within newly laid out irregular fields. There are many references to newly enclosed land in the fourteenth century records of Blackmoor Forest in Dorset. Inge Iram de Berger, who held large estates in the region, was granted permission in 1314 to 'reduce to cultivation' two separate areas of land of 76.5 acres (31 hectares) and 108 acres (43.7 hectares) in the parish of Hermitage. Further examples from the Cotswolds and the wooded lowlands of Shropshire tell the same story. This is also indicated in the modern pattern of irregularly shaped fields, often bounded by thick and botanically rich hedges.[22]

In the uplands of northern England and Wales, permanent farms developed out of temporary sheilings (or summer grazings), a change that is very difficult to date from the archaeological evidence.[23]

Existing villages expanded, often beyond recognition, either by infill, linking up originally separated nuclei, or by new planned extensions. This included settlement along the edge of village greens, indicating that the green, with its communal grazing rights, was becoming central to the village economy. As the population expanded and the conversion of waste into arable intensified, the value of the remaining commons increased and small farmsteads sprang up along their edges. These were not the houses of landless labourers, as the substantial houses often still standing on the same site indicate. They provide landscape evidence for the line of the old common edge in areas where it has long since been enclosed.

The creation of new fields by assarting is often indicated in the documents which become far more available in the bureaucratic period which followed the establishment of the Norman kingdom than in the Saxon period before. In Wheldrake, Yorkshire, we know that the village fields doubled in size between 1150 and 1235, and this is no isolated example. The new fields and their furlongs were sometimes laid out in a very regimented way, which suggests a degree of central control, while in other instances we know that this assarting was a communal effort.[24]

At Hanbury, a village between Droitwich and Bromsgrove, the population doubled between 1170 and 1299. This resulted in the proliferation of small manors, some with farms. By 1240, there were eight separate sub-manors. This expansion was achieved by increasing the intensity of use of existing resources and by clearance of new land, particularly in the thirteenth century.[25]

In the Wealden parish of Laughton it was again the thirteenth century which saw the greatest period of assarting. The main outlines of the local landscape had been established by 1325 and there was little change before 1530, when Sir William Pelham took 1,200 acres (485 hectares) in from the common and waste for demesne woodland and park – a move which, not surprisingly aroused much opposition.[26]

An important change in the law at this time was the Statute of Merton of 1236. This made the taking into cultivation of the common land a far easier business. It allowed the chief tenant (in practice the lord of the manor) to assart land for the exclusive use of himself and his tenants, provided that sufficient common was left for the needs of the village community. This meant that after 1236, it was recognized that the lord of the manor was the ultimate owner of the common, but the rights of the free tenants had to be recognized. Where there were free tenants the commons were more likely to survive.[2]

The problem of farming the outlying land from a nucleated village in the centre may be one of the reasons for the setting up of isolated farms by individuals colonizing new areas.

In the heavy clay woodlands such as Arden and the Weald this involved reclaiming land probably first cleared in the Iron Age and Roman times. The hamlet-type settlements of the Weald represent frontier communities, whilst in the forest of Arden hundreds of moats were dug to provide well-drained house platforms. These dispersed farms, once established, were farmed by families or individuals rather than communities. The contrasts between the settlement patterns of the isolated farms in the old forest areas and the nucleated villages of the nearby open field regions is still a striking one.

Colonization also took place up the sides of the Yorkshire and Pennine valleys where marginal land was forced into cultivation. In the Lake District isolated farms, high up on fell slopes with stone walled paddocks and easy access to extensive high grazing were first recorded between 1100 and 1300.

Isolated farms can, however, have several origins: either of a 'relic-type', the surviving farm in a deserted village, a 'secondary' settlement created as part of the colonization of new land; or simply a survivor of the older local settlement pattern which was never replaced by nucleated settlements.

The buildings of farming

Although there is much in the landscape today to indicate farming practice in the early Middle Ages, little survives in the way of buildings. Those remaining are usually on monastic estates, built to the highest of standards. Some large manorial establishments, too, maintained large farms in which specialist activities required their own buildings. The Cuxham manor records from the early fourteenth century describe a set of buildings grouped around a yard. There were two barns, one for wheat and the other for 'spring corn', a carthouse, hen-house, pigsty, byre, stable, granary and dovecot as well as a drying kiln used for peas, vetch and possibly malting. Twelfth and thirteenth century leases for manors owned by St Paul's, London, describe aisled barns as well as other buildings. For Kensworth in Essex, more detail is given. The oxhouse was 33 by 12 feet and 13 feet high (10 × 3.5 × 4 metres), the sheep house 39 by 12 feet and 12 feet high (12 × 3.5 × 3.5 metres) and the lamb house 24 by 12 feet and 12 feet high (7 × 3.5 × 3.5 metres). At Chingford, again in Essex, it was necessary to go through two outer courts of farm buildings to reach the main house. It is clear therefore that on the larger manorial farms there was a wide variety of buildings, many of which may have been of flimsy construction.[27]

The individual small farmer made do with far less. Over much of upland Britain the longhouse, where animals lived at one end and the farmer and his family shared the same front door, living at the other end, was typical. Away from areas of traditional stone building little is known of the farm buildings of the small farmer before the fourteenth century. There does not seem to have been a longhouse tradition in these lowland areas and so presumably a group of buildings beside the street or green, possibly around a yard, would have been usual.

It is clear that by the fourteenth century many of the characteristics of the farming landscape which divide the country into its distinctive agricultural regions were firmly in place; the isolated upland farms of the north and west, and those, often with moats around them, found in the old forest areas of the clay belts, the nucleated villages of the open-field regions and the green villages of the pastoral north-east. The huge expanses of ridge and furrow divided into furlongs across the arable regions were fully exploited by 1300. The small, irregular hedged fields were typical of the old woodland areas and walled enclosures of the uplands. The different forms of farm buildings too were already established in their regions; the longhouse tradition of the north and west carried on into the early modern period, whilst elsewhere a farm house with a separate barn, stables and other outhouses arranged more or less around a yard were already typical.

Summary

The period of medieval expansion is one from which more is likely to remain, particularly on marginal land which has not been cultivated again until recently. The monasteries built barns and farms high up on the moors whilst the medieval peasant built walls. Field patterns, either surviving, or on early editions of the Ordnance Survey maps may show field boundaries which suggest nibbling along the edges of commons or moors resulting from assarting. Moats are also often an indication of a well-to-do farmer colonizing new land.

As more buildings either survive above ground, or have been found in excavations, it is possible to recognize regional differences between both aisle and cruck construction and between the longhouse and courtyard plan of farmsteads.

5 Contraction and consolidation - the late medieval landscape

The Black Death

The early fourteenth century saw a halt in the steady agricultural expansion of the previous 800 years. There was a gradual deterioration in climatic conditions. Crop failure, which had previously amounted to no more than one season in twenty, increased in marginal areas to one in three. Yields too were declining as the fertility of land, continuously cropped for a thousand years, was gradually sapped. The years 1315 to 1317 saw disastrous harvest failures. The population was outstripping food supply and so was vulnerable to the bubonic plague when it struck at its most virulent in 1348, and then again in 1361, 1369 and 1374. The population dropped by between a third and a half and did not reach its 1348 level again until 1600. The fact that children and young people were particularly vulnerable meant that the effects of the disaster were felt for many generations to come.

On many estates the number of tenants went down by half. At Forncett in Norfolk for instance, the organization of the manor and the demesne was revolutionized as a result. By 1376 many of the demesne buildings had fallen into decay and some new ones were being erected. There was no resident bailiff and instead the estate was visited briefly if regularly by a travelling manager who arranged for the buying of sheep and their housing in the courtyard in the winter.[1]

The decline in population meant it was far more difficult for landlords to exact labour dues from their tenants. The number from whom weekly work at Forncett could be demanded during thirty-five weeks of the year had reduced from twenty-five to seven. Some of those no longer owing labour services had commuted them for rent and one tenement had reverted 'for lack of tenants'. At the same time wages were rising and so it is not surprising that the demesne was being leased out rather than being farmed by the lord working it with hired labour. Evidence from the estates of the Bishop of Worcester, as well as Forncett, show that both existing peasant tenants and landless serfs were able to take advantage of the situation and take on land. Peasants enlarged their farms by taking over the holdings of deceased neighbours or renting parts of the demesne. The majority of the holdings of former serfs at Forncett were small, between 1.5 and 3 acres (0.6 and 1.2 hectares), but a few held considerable acreages.

At the same time rents were declining. On the Bishop of Worcester's land at Bibury the drop of income from rent was from over £11 in 1299 to between £7 and £8 in 1370. Two thirds of his land in the village of Hanbury, Worcestershire, was vacant in 1349.[2]

The effects of these changes on the farming landscape were considerable. First, there was the retreat of corn from the margins of cultivation as the arable was either abandoned or converted to grazing.

The proportion of pasture to arable in Warwickshire went up by a third.

This change had begun before the Black Death. At least 5,000 acres (2,024 hectares) in Cambridgeshire had gone out of cultivation in the five years before 1342. Flooding was resulting in the abandonment of fen. But also the general economic downturn, as well as a decline in fertility, had resulted in land in both marginal and good soils across the county being abandoned.[3] The Black Death came on top of the previous difficulties and resulted in an increased rate of desertions. In the years after 1350 the grain production of the Huntingdonshire manors of Ramsey Abbey fell by a half. In areas of upland such as the high chalk downland of Dorset, ridge and furrow survives on land which was not cultivated again until the war crises of the nineteenth and twentieth centuries. Instead, more sheep were being kept and huge flocks of between 1,000 and 7,000 are recorded, particularly on monastic lands. Oats was being grown at 1,000 feet (300 metres) in the early fourteenth century in the Derbyshire Peak District, but not later: an indication of the pressure on land before the Black Death and the retreat of arable thereafter. Strip lynchets on the hills of Somerset, for instance, were no longer used for grain. Land became derelict in a number of south-east Gloucestershire villages in the fourteenth century and remote settlements were abandoned. This was accompanied by a great increase in sheep. Somerset and Wiltshire were the leading counties for cloth production at the time. Glastonbury Abbey had 1,504 sheep in 1330. Most of Mendip and the Quantocks was unenclosed and simply used as huge sheep walks. On the estates of the Bishop of Worcester there was a drastic reduction in the number of oxen kept for ploughing, and a corresponding increase in the number of sheep.

On Dartmoor, arable farming arrived late and was abandoned after a short time. At Holme Moor the cultivated fields had reverted to pasture in the late fourteenth century.[4] Similarly, Hound Tor is a settlement associated with an expansion of cereal cultivation in the early thirteenth century whilst both settlement and cultivation retreated to the lower slopes a century later.[5]

Deserted medieval villages

As marginal land was abandoned, so often were the villages associated with it. Nearly 3,000 deserted villages have been identified in England. Most disappeared as a result of slow change beginning with the Black Death, and final desertions tended to be in the fifteenth rather than the fourteenth century. There was little desertion as a direct result of the Black Death, but rather a weakening of the village community and the communal farming system. Far from evicting tenants, landlords tried to maintain their farming systems much as before, but with decreasing acreages. Villages in the marginal areas such as the sandy Brecklands and heavy clays of East Anglia, and the Yorkshire and Lincolnshire Wolds were particularly vulnerable.

In lowland areas too, it was often the poorer land that was abandoned first. Grenstein, a deserted village in Norfolk, was settled late. It is surrounded by shrunken and deserted villages, but its nearest neighbour, Mileham, was first settled in the middle-Saxon period. Mileham lies in a sheltered position, partly on well-drained soil, and was a successful village in the Middle Ages. Grenstein on the other hand was on heavy boulder clay on an exposed windswept hillside. At its most populous it contained twenty-six crofts and was probably deserted by the fifteenth century.[6]

However, desertion was not simply an abandonment at the margins of cultivation. Assarts of the twelfth and thirteenth

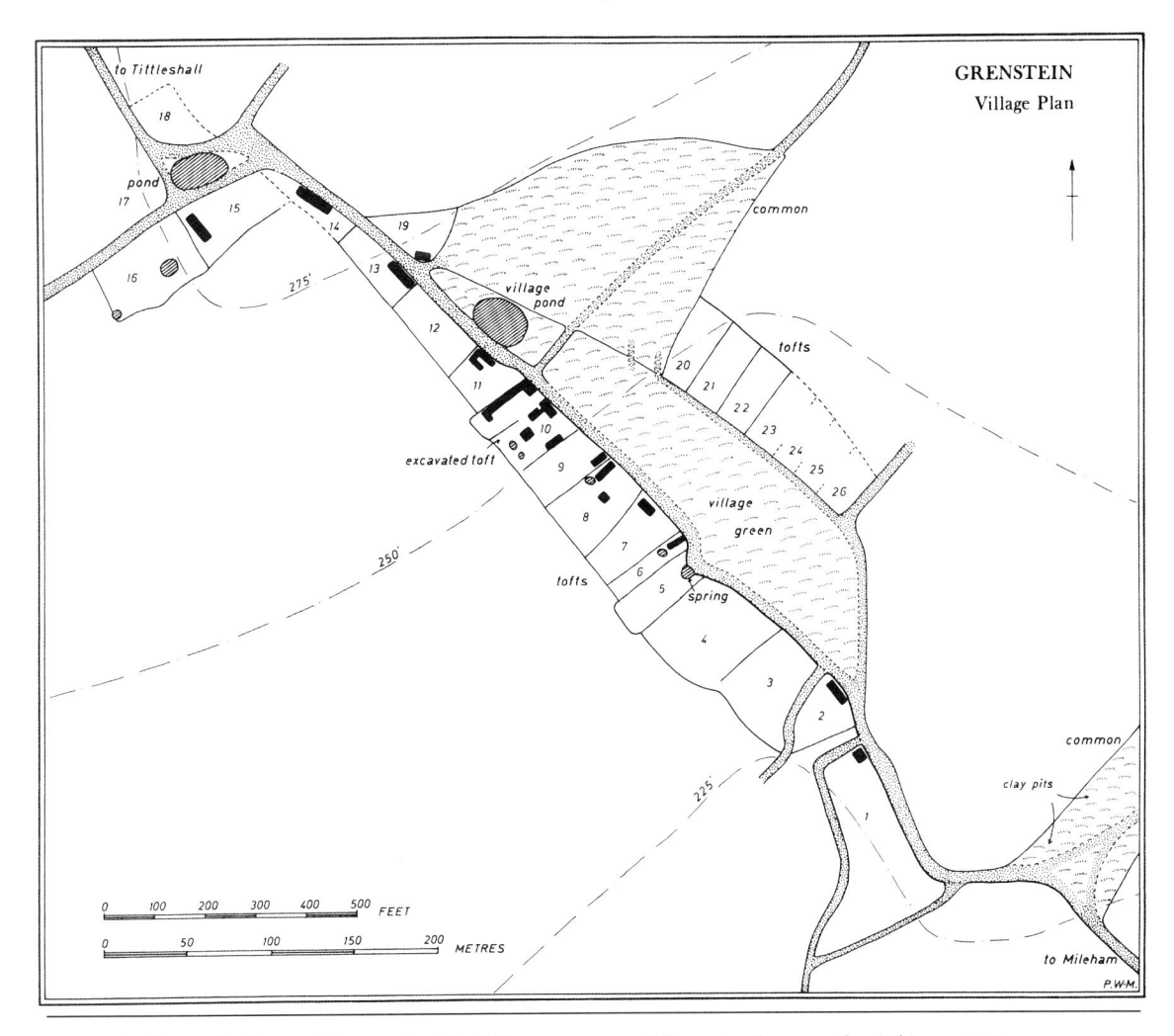

Contraction and consolidation – the late medieval landscape

27 *The deserted village of Grenstein, Norfolk, reconstructed from air photographs. This post-Norman settlement was founded on marginal land and its twenty-six farms along the green edge were gradually abandoned during the late Middle Ages (Field Archeology Division, Norfolk Museums Service)*

centuries had often been made by free tenants who were tenacious in keeping their holdings in adverse conditions. On the manor of Havering in Essex for instance, extensive assarting had pushed the frontiers of cultivation onto unrewarding soils and these settlements, perhaps because of the proximity of the London market, survived the fourteenth and fifteenth centuries.

Much more common than the totally deserted village, and indicative of population decline, are shrunken ones. Here there are areas of earthworks and house sites, either in vacant plots between surviving houses or on the outskirts of the villages. When an area is studied in detail, the relationships between deserted, shrunken and flourishing villages and the changes in farming that this implies can be seen.

In the Worcestershire parish of Hanbury, earthwork surveys, combined with the study of pottery-scatters and documentary evidence, have made it possible to locate eighty deserted 'messuages', or individual farmsteads. Their locations indicate that they were not necessarily the last settled, or on the most marginal land. A typical messuage, as revealed by the earthworks, was usually inhabited by a single family and stood alone beside a road or lane. It was about 33 by 87 yards (30 × 80 metres) and was surrounded by a bank or ditch. Inside were rectangular platforms suggesting the presence of two or three buildings: a dwelling, a barn and possibly a bakehouse or livestock shed. Sunken areas mark the sites of yards with perhaps a garden plot at the rear. Beyond the boundary ditch was the ridge and furrow.[7]

In the modern parish of Mudford, just to the north of Yeovil, Somerset, there were eight separate settlements in the thirteenth century, although one was probably never more than an isolated farm. The other seven were hamlets, with a church at Mudford. Of these, two have disappeared, four are severely shrunken with extensive earthworks, and only Mudford is reasonably intact. There is no documentary evidence to date this desertion, but typically it would have been a gradual decline until the late sixteenth century when the open fields were enclosed and land realloted.[8]

The parish of dispersed settlements at Hartland in Devon has similarly been studied using medieval rentals, unusually listing not only the tenants, but their farms and hamlets. In 1301 there were some isolated dwellings, but most scattered settlement sites consisted of between two and four farms. A few sites were occupied by over ten farmsteads. Between 1350 and 1450, however, nine sites were deserted and many hamlets dwindled to become the isolated farms which they are today.

The richness of the fen-edge in Cambridgeshire allowed the population to grow in the early Middle Ages to levels that were not recorded again until the nineteenth century when more land was brought into cultivation. The arable at Landbeach doubled between 1279 and 1316. However, the Black Death wiped out up to half the population in many of these fen-edge communities resulting in many uncultivated plots in the open fields. The manorial records show the ceaseless struggle of the lord's officers against dilapidation and waste. In spite of their efforts, arable land reverted to fen.[9]

It was not only in the areas of nucleated villages that fields and settlements were abandoned. In areas of dispersed settlement, such as east Suffolk, it was the outlying parts of villages that were abandoned and many small hamlets and isolated farmsteads disappeared. In some areas of the uplands and woodlands a loss of between a half and a third of the isolated farmsteads and hamlets was experienced.

Hamlets and isolated farms often predate villages and in some ways represent the more normal form of agriculture, before population pressure forced a change to communal farming. Their importance must therefore not be overlooked. For instance, a map of deserted sites on Exmoor shows the importance of farmsteads rather than villages in the uplands. The vast majority of deserted sites within Somerset were hamlets rather than villages. This includes 100 farmsteads on Exmoor where twenty per cent of the farm sites known in the fourteenth century are now deserted.[10]

A changing farming system

It would be a mistake to assume that conversion of arable to pasture was necessarily a retrogressive step, particularly when there was a smaller population to feed. Rather than collapsing, farming was restructured. It could certainly be very profitable to those who converted to it as the fifteenth century 'wool churches' of East Anglia and the Cotswolds show. In Warwickshire it was often those from ranks lower than the gentry, the 'yeoman graziers' who benefitted most. Before 1349 tenants with more than 30 acres (12 hectares) were rare, with units over 100 acres (40 hectares) being in the hands of landlords, but by the early sixteenth century they were mostly held by tenants, some of whom were tenants exploiting specialist pastoral farms.

The emergence of a farming elite

For those who survived the Black Death, and their successors, there was the opportunity to rise up the social ladder. Firstly, landless people took over the vacant holdings, and this was followed by amalgamations. Calculations for West Country manors show the average size of holdings increasing from 22 to 47 acres (9 to 19 hectares). It was these amalgamations by tenants, rather than depopulations by landlords, which led to a decline in the number and size of settlements.

This general population decline resulted in many changes in farming patterns which in turn affected the landscape. The more enterprising tenants took advantage of the situation to expand their holdings in the open fields. Many of these farms now included 60–80 acres (24–32 hectares), whilst those of their grandfathers would have contained no more than 30 acres (12 hectares). The strips were often consolidated in their individual furlongs, which were then enclosed with hedges, thus preserving some of the boundaries of the earlier system. At Chippenham in Cambridgeshire individual farmers were accumulating the 15 acre-units (6 hectares) which documents show had comprised a single farm before. In 1384 John Lenote took two 15 acre- (6 hectare-) holdings, another in 1392, 1399 and 1400. The lord of the manor tried to prevent these all becoming consolidated in one block of land, but failed. Communal farming was thus abandoned and sometimes the new farmers moved out of the old villages, building single farmsteads, often enclosed by a moat. There were thirty-seven households in the parish of Clare in the Chilterns in 1279, but by the first surviving map in the 1630s, the fields had been reorganized around four isolated farms.[11]

Associated with the amalgamation and exchange of strips as well as a shift to pasture, enclosure across the country was slow but steady. By 1500 the amount of enclosed arable probably equalled that still cultivated in the common fields.

Brookend, on the Gloucestershire border of Oxfordshire, is an example of an area of late colonization initiated by the Abbey at Eynsham. By 1279 there were sixteen households and the land was fully exploited. 700 acres (283 hectares) were sown each year with cereals and pulses while a few sheep were also kept. The village survived the Black Death and the reason for its desertion in the mid-fifteenth century was economic rather than demographic. Land was being consolidated in the hands of a few tenants. Despite the efforts of the Abbey, the village was changing from a community of peasant holdings to a collection of yeoman farmers. The soil was also becoming impoverished after 250 years of intensive farming. Larger farms could afford a more extensive system of farming, involving the keeping of more animals to replenish fertility.[12]

At the same time as the peasants were increasing their holdings, the lords were abandoning their demesne land. As labour services were becoming more difficult to enforce, and were being replaced by wages, farming by the landowner on his own behalf became less profitable and so his land was let. Demesne farming had all but disappeared by 1450 and the new class of independent yeoman farmers were well able to take advantage of the situation.

The growth of the home-based textile industry meant that after 1450 much of the enclosure taking place was for sheep pasture.

These fields tended to be larger than those for arable, but they might still follow the lines of the old furlongs. This was particularly true in the midlands where a third of Leicestershire was converted to pasture between 1349 and 1520. The sheep farmers often operated on a very large scale and the foldcourse system, typical of the light lands of East Anglia, developed to meet the needs of the landlords. Tenants were not allowed to keep sheep, but their landlords had the right to graze their flocks across the open fields of the tenants during the winter months. This allowed for the manuring of the fields, but was resented as it prevented enclosure. However, in general, the increase in stock after the Black Death had a beneficial affect on crop yields as is clear from the ecclesiastical accounts of the Bishops of Winchester.

By the early 1500s, the population was increasing again which meant renewed pressure for land and therefore resentment of enclosure, resulting in the unrest of the early Tudor period.

28 *Reconstruction of the village of Wharram Percy as it was in its prime in the early fourteenth century, with the manor house in the foreground (English Heritage)*

The buildings of the late-medieval farmer

Buildings form an intergral part of the farming landscape and it is from the period following the Black Death that they begin to survive in any number.

Manorial records, both for secular and ecclesiastical estates, make it clear that as well as barns, the larger farmsteads included granaries, stables, livestock housing, cartsheds, dovecots and malt houses.

This type of farmstead with the buildings arranged, albeit haphazardly, around a yard is in marked contrast to the single longhouse more typically found on the smaller peasant farms of the north and west. Although found over a wider area of western Britain in the Middle Ages, by the sixteenth century, longhouses were confined to the north and west, from Devon to the Lake District. In its most primitive form, as revealed in excavations at West Whelpington, Northumberland, Wharram Percy, Yorkshire and Hound Tor, Dartmoor, the occupants were only separated by a paved way, although this was later replaced by a partition. Accommodation for the farmer and a small number of stock and a few chickens was provided under one roof, but there was no provision for cereal storage or threshing. The small size and emphasis on livestock means that they were usually associated with small-scale peasant farming in the livestock areas of the country. But, that this arrangement was not restricted to the poorer farmers is shown by the very substantial examples of farmsteads, with finely moulded doorways and windows, which survive on Dartmoor, in Wales and the Lake District.

By the fifteenth century, this contrast between the courtyard layout of the lowlands and the linear farmsteads of the uplands is clear. The linear arrangement was more suited to the smaller upland farms and buildings on sloping ground. The cattle end

29 *A Dartmoor longhouse at Oldsbriun, Poundsgate, Widdowcombe-in-the-Moor, Devon (Peter Beacham)*

was always downhill from the living accommodation to help with drainage.

In the lowlands, longhouses were never typical, even on the smaller farms. Excavations on deserted village sites reveal individual courtyard tofts even on the simplest farmsteads. The divide between longhouse farmsteads and those arranged around courtyards is not one associated with size or prosperity of farms, but with regional differences.

Other types of farm building showing distinct regional distributions are aisled and cruck-framed buildings. Building huge barns with a wide roof span could only be achieved by inserting parallel rows of supports holding up an aisled structure, and so where cereal output was particularly important, and the estate was large enough, great aisled barns were built. They are most likely to be found on monastic estates where, as we have seen, some date back to the thirteenth century. A later group were built in the West Riding of Yorkshire and east Lancashire in a period of high grain prices and improved climate in the sixteenth and seventeenth centuries. Some are on farms over 800 feet (245 metres) above sea

30 *Reconstruction of the interior of a cruck barn, Wharram Percy, Yorkshire (English Heritage)*

level which would now be considered marginal for cereal growing.[13]

This type of structure is not found west of the Cotswolds, whilst cruck buildings are typical of the midlands and north-west of the British Isles. Many crucks, particularly in barns, are massive structures and it is their strength and size which explains their survival. The curving upper timbers make it difficult to insert an upper floor, but they were ideal for barns, and of the 3,000 recorded cruck buildings in 1980, nearly 800 were built for agricultural use, mostly as barns. They are often very difficult to date as this method of construction is known from the early Middle Ages and remained in use in remote areas until the nineteenth century.[14]

The buildings, landscape and organisation of farming was very different in 1500 from the situation at the height of medieval prosperity in 1300. The tight nucleated villages now often had gaps in them where tofts had been deserted and small upland hamlets had been reduced to isolated farms, whilst in contrast the isolated farms of the new independent farmers stood out amongst their newly enclosed fields.

The communal system on which the intense cultivation of the early Middle Ages had relied was now less necessary. A new class of independent farmers' enclosed groups of furlongs in the open fields, and a haphazard patchwork of hedged fields, sometimes interspersed with narrow strips, became a familiar pattern over much of the lowland landscape. The new yeoman farmers were better off than their peasant predecessors and so could afford to erect better quality houses and farm buildings, some of which, particularly barns, survive today. Many exhibited their wealth in the building of fine parish churches, typical of the so-called wool villages of the Cotswolds and East Anglia.

By 1500 the countryside reflected the contrasts that were developing within society. Whilst the great barns and manor houses of the yeoman farmers do sometimes survive, the peasant tofts are only to be found in excavations. At Wharram Percy, a deserted village, the site is open to the public. The outlines of the excavated buildings are marked on the ground, the significance of the surrounding earthworks explained, and the ruins of the two remaining buildings consolidated. The working of a late medieval village within its fields can be appreciated in a unique way. The early sixteenth century landscape was already exhibiting the variety brought about by the greater independence of the later Middle Ages. As the population began to rise in Tudor times, regional patterns in agriculture began to develop. Although all farms had to provide for their own subsistence and so to some extent had to remain mixed, the various regions of the country were beginning to produce the foods to which they were most suited. The growth of towns and local markets meant that the whole country was drawn more firmly than ever into an integrated system of agricultural production.[15]

This in turn meant that as regional specialization grew, so did the importance of the market economy. Farmers became part of an entrepreneurial society which increased the divides between the small subsistence and larger scale commercial farmer.

Summary

The contraction of the population left evidence of the earlier farming systems behind it, such as deserted villages and abandoned ridge and furrow. Although these are evidence of the previous era, their very existence is an indication of the degree of contraction that took place. For those that were left, however, life could be more prosperous and the survival of fine late-medieval manor houses, sometimes with farm buildings, are an indication of this. Surplus wealth, nearly always made in agriculture, was often syphoned off into church improvements.

6 | 1500 - 1700 The rise of the gentry farm

Although the period 1500–1700 was marked by great political upheavals, the dissolution of the monasteries and the Civil War, as well as a major outbreak of plague, it is also 200 years which brought prosperity and progress to many. It was the time when the foundations of an agricultural system and the farming landscape, which we recognize today, were laid.

The early sixteenth century saw the emergence of the yeoman farmer as the feudal system finally collapsed with the accession of the Tudors, in 1485. His rise was helped by two things, firstly, an increase in the price of grain and a steady market for wool, which meant that his income doubled between 1500 and 1640, and secondly, the fluid land market of Tudor and Stuart times in which the dissolution of the monasteries, resulting in the sale of land covering a quarter of England, must be seen as a major part. In the seventeenth century the Stuart kings, also short of money, sold royal forests for enclosure.

Both the independent farmers and the old-established landowners benefitted from these sell-offs, but the landlords saw themselves losing out as agricultural prices rose. Much of their land was let on long leases at low rents, which had been supplemented by high entrance fines paid when a new tenant, usually the son of the previous tenant, took over. Over the 200 years this situation changed. Leases became shorter, rents higher, and entrance fines became much less important, finally disappearing altogether.

In the second part of the period, civil war and plague slowed down the economic and population growth of the country; grain prices fell by twenty-five per cent between 1640 and 1750, whilst those of livestock rose. There was therefore an incentive to improve pasture and look for alternatives to cereals. Market gardening, dairying and orchards became more important near centres of population, whilst alternative crops, such as saffron for dying, were introduced where there was a plentiful labour supply to process this labour-intensive product.

The two contrasting economic situations of high and low prices had similar effects on farming practice. High cereal and wool prices encouraged farmers to increase their output, whilst the later recession intensified the search for improved methods to keep up profits.

Enclosure

One of the most publicized changes that took place in the farming landscape in early Tudor times was enclosure. Enclosure of the common fields had been going on in some areas throughout the Middle Ages, but with the decline in population and the increase in the value of wool, it was enclosure for sheep walks which became a political issue at the end of the fifteenth century. The popular belief was that enclosure was always the work of the landlord against the wishes of the majority, in order to increase his often already very large sheep flock at the expense of others, thus leading to depopulation. The first anti-enclosure act was passed in 1489. This 'husbandry act' made it illegal to cause the decay of a 'house of husbandry' (defined as a farm with more than 20 acres (8 hectares) under the plough). A further act of 1515 made it illegal to convert land from tillage to pasture. Acts of 1533 and 1555 tried to prevent the keeping of large numbers of sheep, and a final anti-enclosure act was passed in 1597. In spite of all the publicity and strong feelings against enclosures incited by the pamphleteers, there were very few prosecutions under these acts, which suggests that enclosure was in fact no longer general.

Of all the cases brought before the Court of the Exchequer between 1518 and 1568, ninety-eight per cent were from the midland counties of Northamptonshire, Oxfordshire, Warwickshire, Buckinghamshire, Berkshire, Nottinghamshire, Leicestershire and Derbyshire. In these areas the grazing system involved a mixture of permanent pasture and long leys with occasional cultivation of grain or fodder crops. This sort of husbandry was only possible in enclosed fields and so it is not surprising that early enclosure is found here, but it is also true that much open field survived. Deserted village evidence suggests that the main period of this type of enclosure was between 1440 and 1520 in villages which were vulnerable because they had already suffered depopulation. John Rous of Warwick listed about seventy-five depopulations mostly in his own county, 'within living memory', in about 1486. However, he does not say how many people were left in the villages before the final clearance. In more than two-thirds of the proceedings against enclosers following the 1517 enquiry, the alleged depopulation involved only one house; in only twelve cases out of 482 was the removal of more than ten houses alleged. Where documentary evidence has enabled the study of the final years of a village, as in the case of Brookend, Oxfordshire, and Hatton Warwickshire, it is clear that depopulation had been a slow and natural process until the landlord had eventually taken the land into pasture.[1]

In Lincolnshire depopulating enclosure may have gone on longer than in the Midlands. A return of 1607 lists 199 farm houses and 140 barns, cottages and stables as being totally decayed and 172 houses of husbandry as being decayed. However, there is no indication of how long they had been derelict.[2]

By the time the acts of parliamant were passed, wholesale enclosure for sheep was probably already declining. Wool prices peaked in 1550 whilst the population was rising and with it the price of grain, reducing the incentive for conversion to sheep farming.

Much more widespread, if less publicized, must have been enclosure in the south and west. By 1600 Kent, Cornwall, Devon, Essex, Cheshire, Lancashire and Monmouthshire were described as 'wholly enclosed'. It may well be that Cornwall and parts of Devon had never contained much open field, but Kent and Essex had been enclosed to provide for the London market. 'Heavily enclosed' counties included Shropshire, Hereford, Sussex, Suffolk, Surrey, Somerset, Northumberland, North Riding of Yorkshire and Dorset. Again some, but not all of these, were serving new expanding urban and industrial markets.

Somerset was described by Norden in 1607 as a county 'prosperous through enclosure'. This is an oversimplification of an immensely complicated and varied situation where both early enclosure and the long survival of open fields are found side by side. The Vale of Taunton was mainly pastoral and probably never an open field area, whilst much of the lower Mendip was enclosed by agreement between 1500 and 1780. The rough uplands remained open until the period of parliamentary enclosure, but lower down enclosure was undertaken to allow for the intensifying of farming to serve the growing urban market of nearby Bristol.

A few villages in the county were almost completely enclosed by agreement by the seventeenth century. At Cricket St Thomas, between Crewkerne and Chard, only one open field survived in 1543, and in 1540 there was an agreement to enclose it. Five arbitrators were appointed for this purpose, one for the lord of the manor, one for the rector and three for the tenants. Many villages disappeared as part of the enclosure process, usually remaining only as single farmsteads.

In Suffolk, much of the area of light soils west of Bury St Edmunds remained in open fields, but the heavy boulderclays were largely enclosed by the sixteenth century. Because the commons were large, the open fields did not need to be used for grazing and so it was relatively easy to reach agreement to enclose. A sixteenth century observer of west Suffolk wrote that farmers 'time out of mind used to sever and divide their copyhold land with ditches, hedges and pales at their pleasure and without licence.'

The Dorset landscape began to take on its present appearance between the late fifteenth and mid-eighteenth century. A farming system specializing in sheep production brought wealth to a large number of people. Former downland was enclosed into large, generally rectangular fields, often with the agreement of all parties affected. Arable was enclosed as a result of slow and piecemeal change, resulting in a far more patchwork layout than on the higher land.

Half of Oxfordshire was hedged by 1730, but less than a third of townships were completely enclosed. It is not easy to pinpoint the early enclosures in the modern landscape as they were created in smallish parcels and so were often refashioned when parliamentary enclosure took place around them. Old hedgerows and dispersed secondary farms are the best indications that survive of areas of pre-eighteenth century enclosure.[3]

William Marshall, writing in 1787, regarded east Norfolk as 'very old-inclosed country' and described the fields as small with high hedges full of trees. This 'abundance of petty inclosures' seemed to Marshall disgraceful in an arable area with some commons and waste still remaining.[4]

The chalk downlands and Cotswolds lost their open commoning rights in the seventeeth and early eighteenth century as land was enclosed, often for arable.

It was the midland belt that was the heartland of enclosure in the sixteenth and seventeenth century. In much of Leicestershire for instance, enclosure was a gradual process over 400 years, with a speeding up of the process between 1607 and 1710 when the percentage of the county enclosed increased from twenty-five per cent to forty-seven per cent. The type of enclosure changed greatly over the period. On the whole, enclosure began in those districts least favourable to arable agriculture, moving into those most favourable at a later date. Before 1550 enclosure was usually the work of the manorial lord who was most likely to be able to act in decayed villages where there were few to oppose him. It is perhaps significant that the villagers themselves did not see enclosure as a solution to their problems, but preferred to contine farming as they had always done, in a communal system. It was only the landlords, operating on a larger scale, who saw the advantages of enclosure, usually for commercial sheep farming.

The period 1550–1580 was relatively

quiet in Leicestershire's enclosure history, but when in became important again at the end of the sixteenth century, circumstances had changed. Sheep farming was not the only profitable way of farming enclosed land. Although the initiative still often came from the landlord, the freeholders were far more willing participants, and by 1800, all but the smallest tenants usually saw the benefits of enclosure.

This slow but steady enclosure stretching back into the Middle Ages meant that by 1600 about forty-seven per cent of open-field England was already enclosed. A further twenty-four per cent was enclosed during the following 100 years, making this the most crucial period in enclosure history when the country swung from being a mainly open to a mainly enclosed landscape.[5]

It is in the sixteenth and seventeenth centuries that we enter the era of maps. The emerging generations of new land owners wanted to know what exactly they owned and who they might have to buy out if they wished to enclose. With the existence of maps the study of the landscape takes on a new dimension and these early maps show that almost every parish in England had at least a few hedges. By comparing these maps with recent ones and with aerial photographs it is clear that the majority of these hedges were still there in the mid-twentieth century.[6]

It is after 1600 that enclosure really began to speed up and its character to change. Instead of being the work of the landlord at the expense of his tenants, it was now more likely to be enclosure by agreement between a group of tenants or small owner-occupiers who saw enclosure as being to their mutual advantage. The details of such improvements are hard to find and it is likely that only when there was a dispute were records made. Occasionally, as at Great Linford in Buckinghamshire, maps survive of the parish before and after enclosure. Here it was the result of an agreement between the land owner, freeholders and copyholders, and the 'inconvenience' of the old system was given as the reason. At Marston in Lincolnshire the reasons are spelt out more specifically. Once enclosed the land could be 'ploughed, tilled or laid to grass as husbandry and profit incite, and not as necessity of neighbourhood compels' as in the open fields. It is clear that this enclosure of 1630 was not primarily to enable conversion to pasture, although there are still some examples that led to depopulation, as at Haselbech in Northamptonshire.

The impact of enclosure on the landscape was considerable. Open arable fields extending over several hundred acres were replaced by closes varying in size from 2–3 acres (about 1 hectare) to over 100 acres (40 hectares) fenced in with ditches and banks on which hedges and trees grew. If the enclosure was piecemeal, the fields could be of very irregular shape, with hedges conforming to either the individual strip boundaries or those of a group of strips which made up a furlong. They might be long and narrow, or still retain the curved 'S' shape of the old strip. In these cases the hedges often form a sinuous or curving network of mixed hedges.[7] In spite of the hedge removal of recent years, these irregular fields can still be seen in areas of old enclosure, or picked up on the early editions of large-scale maps.

Enclosure by agreement often allowed for the relaying out of the fields with total disregard for the open-field layout. The result could be a landscape similar to that created by the later parliamentary enclosures, and the best way to distinguish between the two is by the variety of species in the hedge. Whilst the early ones have become species rich, the parliamentary ones are still predominantly hawthorn.

The reasons for the increase in enclosure during this period are many, but some of them were the active land market, particularly after the dissolution of the

31 *A view of Swaledale Yorkshire showing the small stone-walled fields with their field barns; almost one to a field*

monasteries, and the increase in the size of estates, making it easier for the landlord to impose his will. Alongside this must be seen the increasing urban and industrial market, a growing population, and a higher standard of living, all helping to encourage an expansion of agricultural production.

Alongside the enclosing of the common fields went the felling of woodland and the bringing of waste into cultivation. In many areas, however, the commons remained open for another 100 years and it was only the open fields that disappeared.

New crops and improved farming practice

It is becoming clear that where the soil was suitable, a variety of new crops could be grown within the open fields, but enclosure meant land improvements, such as under-draining or claying, could be carried out, allowing the new crops to be grown on the heavy soils. Industrial crops such as woad, saffron and madder, or fodder crops such as turnips or improved grasses, were introduced. Flexibility allowed for experimentation with new rotations. A generation of literate farmers were reading widely from the growing number of agricultural books available. Some had visited Holland during times of religious difficulty, or as refugees during the Commonwealth, and brought back new ideas. As grain prices fell after 1640 interest in livestock and particularly cattle increased. The wet summers of the 1640s in which hay rotted in the fields were an incentive to try new fodder crops. Clover and sainfoin were increasingly grown from the 1660s. Turnips were grown in East Anglia by the 1650s, but were not a regular part of rotations until the 1720s. Although we know that experimental farming was carried out in open fields, it was obviously far easier to innovate in enclosed fields and open-field farmers rarely adopted the new methods to the same extent as those with enclosed fields.[8]

Some enclosure, particularly on the chalk downs, was still mainly for sheep. Vast flocks were kept in Dorset in large, somewhat irregular fields bounded by quickset hedges on low banks. The main problem for the flock masters was finding enough feed in the early spring, and one solution widely practised in the chalk valleys of Dorset, Hampshire and Wiltshire from the early seventeenth century was the water meadow. Although few are still in use, they are clearly visible in many a valley. One of the most extensive is on the river Avon at Britford, near Salisbury, where five major channels are fed from the river. They flow on the top of low ridges, which in turn discharge water into smaller channels about 17 yards (15 metres) apart. From these water floods across the fields and is collected again in a further set of ditches and returned to the river.[9]

The work involved in setting up and maintaining such a system was considerable. A main leat was dug from a high point in the river along the top of a series of meadows, and from this feeders were dug across the field back towards the river. A sluice could be opened to allow water to flow along the leat and into the ditches, flooding the fields. The meadows would thus be 'drowned' in the early spring. The moisture and the slightly warmer temperature below the water allowed for the early growth of grass.

32 *Water meadows, Nadder valley, Wiltshire (University of Cambridge Air Photograph Collection)*

The water had to be kept flowing to prevent the grass rotting. The field would then be drained, providing an early bite of grass for sheep from February to May, by which time the grass elsewhere was coming on. The grass could then be flooded again and a hay crop taken. Water meadows such as this were to be found in the valleys of the chalk downs across the southern counties. The earliest documentary reference to them in Dorset was in 1629 at Puddletown, where after a great deal of discussion in the manor court, it was agreed to begin construction. The cost of their construction meant that in the seventeenth and eighteenth century they were confined to the areas where they were of the greatest advantage, such as the southern chalklands where the winters were mild enough for the system to encourage early growth. The fact that a certain amount of silt was deposited along with the water helped maintain fertility and encouraged heavy hay crops. Though no longer used, their ridges, channels, drains and elaborate sluices are still a marked feature of large areas of the valley floors of the rivers Frome and Piddle, as well as the smaller chalkland valleys.[10]

Fenland drainage

The most dramatic example of the spirit of improvement which was behind much of the felling of forest and enclosure of waste carried out in the seventeenth century was the drainage of the southern fenland. Until the 1530s much of this area was owned by monasteries, of which Crowland and Thorney were the most important. The great abbeys had been responsible for the drainage and much of the land had been used for pasture. In few other parts of England did the dissolution of the monasteries have more momentous results for the farming landscape. Huge ecclesiastical estates were subdivided, which led to the neglect of the drains. Henry VIII then set up the Commissioners of Sewers to deal with the many complaints, and there is plenty of evidence to show that areas, such as that around Thorney, which had been 'dry and firm' were now surrounded by water.

By the end of the sixteenth century interest in a grand scheme to improve the drainage was increasing. Some of the new owners of the old abbey estates, most notably Francis, fourth Earl of Bedford, who owned 20,000 acres (8,000 hectares) at Thorney, had money made in commerce or the law and were prepared to risk it in land reclamation.

In 1630 the Earl of Bedford, joined the next year by thirteen other wealthy businessmen, agreed to drain the whole 95,000 acres (38,000 hectares) of southern fenland. The main object was to create summer-grazing grounds, although some land might be suitable for arable. The Dutch engineer, Vermuyden, was employed and his method was that advocated by improvers before him; to straighten the river channels and thus speed up the journey of the water to the sea. His first undertaking, completed in 1637, was 'one of the major engineering achievements of this country' and involved the digging of a straight channel for the Bedford river. It was 70 feet (21 metres) wide and 21 miles (34 kilometres) long, taking the waters of the Ouse to the Wash. This was followed by the digging of other straight channels, but the whole project was interrupted by the Civil War. Between 1649 and 1653 the New Bedford River, parallel to the old, was dug and the area could now be said to be effectively drained. An era of fenland prosperity began. Vermuyden had not, however, foreseen that the draining of the peat would result in its shrinking, and so by the end of the seventeenth century, flooding was again a problem, only partially solved by the building of windmills, and later by steam pumps. The major landscape features of the fens are therefore seventeenth century, but constant improvements and recutting were needed. Enclosure followed

33 *Old and New Bedford Rivers looking north-east across the fens. The old river, to the left, was straightened, whilst the new one was a completely new cut. The area between them (the Wash) is pasture which acts as a holding area for flood water in the winter (University of Cambridge Air Photograph Collection)*

swiftly after drainage, and the variously orientated groups of fields to be found within the fens indicate the different periods of drainage and improvement.[11]

The built environment

The landscape of early enclosure and more dramatically of fen drainage and forest clearance, is not the only indication of the prosperity and spirit of improvement in seventeenth-century rural England. For the first time we begin to find farm houses and farm buildings (mainly, but not entirely, barns), surviving on farms other than those owned by the church or the greatest landowners. The Tudor period saw a great improvement in the standards of comfort, firstly of the houses of the wealthiest, but soon filtering down the social scale. Chimneys and fire places were inserted, upper floors put into open-hall houses and larger windows with glass were added. Brick was replacing wattle and daub in the lowlands, whilst stone remained important where it was available. As well as being more comfortable, houses became larger. Inventory evidence shows the average size of a house in Kent in 1580 being three to five rooms whilst thirty years later it was six to seven. In Essex and Kent medieval houses were being modernized, whilst further from London there is more evidence for new

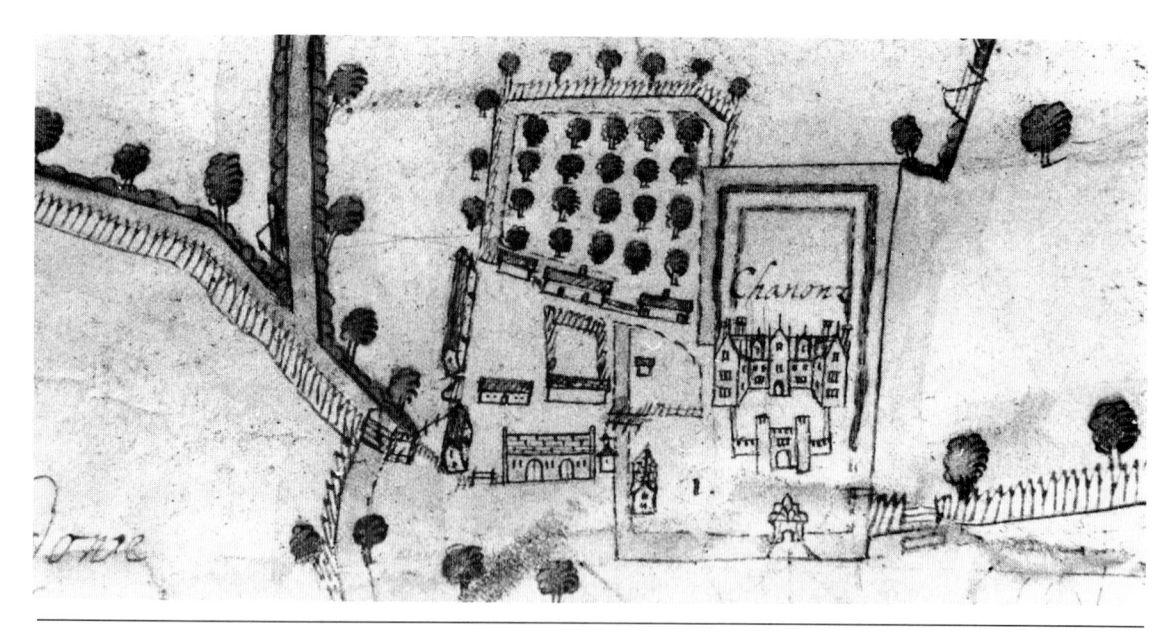

34 *Channonz Hall, Tibbenham, Norfolk, drawn in 1640 and showing a fine seventeenth-century house beside the moat where the medieval house would have been. The farm buildings stand where they would always have been – at the entrance to the moat (Norfolk Record Office)*

35 *Cressingham Manor, Norfolk, showing the farm buildings at the entrance to the moat. These include a barn, stable and granary (Cressingham PCC)*

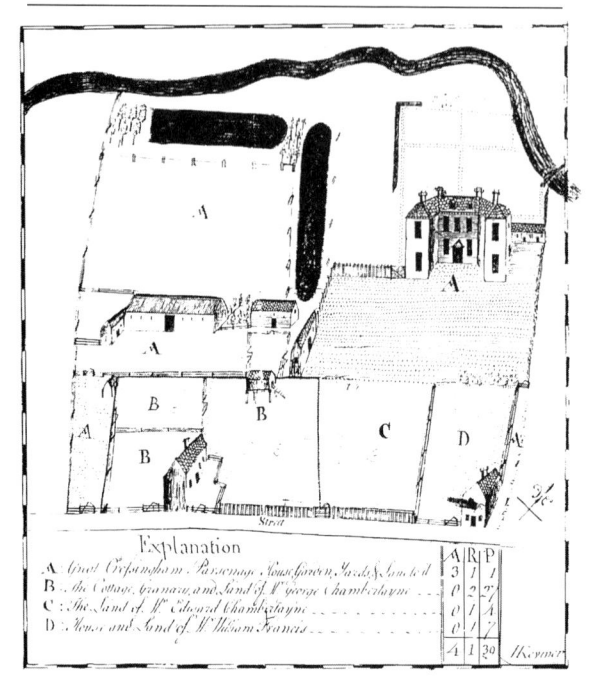

building, especially amongst the yeomen and other successful members of the agricultural community. The highland zone was less prosperous so there was less building in the late sixteenth century. Much of this area was 100 years behind the south, where in the more advanced regions, the family farm was already giving way to enterprises on a larger scale, conducted with paid labour. This change gathered pace in the years after the Civil War and with it the gulf between the rich and poor increased.

In Devon there was a tradition of long and secure tenancies which encouraged building and so the prosperity of the sixteenth and seventeenth centuries has resulted in a legacy of fine farm houses. Some which appear to

36a,b,c,d *This impressive group of buildings on the fertile soils on the edge of the Norfolk Broads includes a fine seventeenth-century house, early eighteenth-century granary, stables and barn*

1500–1700 The rise of the gentry farm

be seventeenth century from the outside are in fact much older, but were refaced so as to appear fashionable and up-to-date. In spite of the increasing standards of comfort, traditional forms were retained, such as the longhouses of Dartmoor. In these houses animals and humans often shared a single door, yet it is clear from the high standards of carpentry, masonry and good decorative detail, that their owners had as high a social status as the farmers in the more hospitable lowland terrains. Longhouses continued to be built in the traditional manner well into the eighteenth century.[12]

The same is true of the other area of England where longhouses survive in any number; the Lake District. Here, unlike Devon, the farmers were in the habit of dating their houses and the large number with seventeenth century datestones is an indication of the wealth and social prestige of the so-called 'statesmen farmers' of this period. This group had gained from the dissolution of the monasteries, and while wool prices remained high, they prospered. Their farms with their fine houses and groups of buildings, often including a two-storey bank barn and 'spinning gallery', can be seen scattered along many a valley bottom. Bank barns, which are to be found throughout the uplands of western Britain, first make their appearance in the seventeenth century. They were built into the hillside and could be entered at two levels. The cattle were stalled at the lower level whilst above was a threshing barn. The purpose of the so-called 'spinning galleries' which stretch the length of a barn, usually adjacent to the house at upper level, is unknown, but they certainly are a most attractive feature.

In east Wales too, datestones begin to appear in the late seventeenth century, showing that this wave of building extended into the Principality.[14] Other regional types of building include the lath house, found in Lancashire and west Yorkshire from the 1650s. Unlike the longhouse, there was a separate entrance for animals and above the cow byre there was a substantial hay loft (or hay mow) for storage of essential winter fodder.

In the border regions of Northumberland, fortified bastel, or peel, houses were built. Here the animals were housed downstairs and the farmer and his family, for safety, lived above. In areas where cattle were widely kept and so hay was important for winter fodder linhays were built from the sixteenth century. These two-storey open-fronted buildings with shelter sheds for beef cattle below, and storage for hay above, were built throughout the west Midlands, south Wales, Devon and Cornwall. By the seventeenth century the isolated field barns, which are a familiar part of the landcape of much of Yorkshire and the valleys of the Peak district, were being built. Here the cattle were kept and fed on the hay collected from the surrounding fields. The manure produced through the winter could be spread on the nearby fields, thus reducing the need for carting both the feed to, and the manure from, the distant farmstead.

Whilst it is now accepted that what Hoskins called 'the great rebuild' of Tudor times was much more drawn out and contained more regional variations than he at first recognized, the era of prosperity for the independent farmers of the sixteenth and seventeenth century can certainly still be seen in their buildings. However, the scale of decay of some of the earlier farms that accompanied the conversion of arable to pasture must not be forgotten. The only evidence for this is often the earthworks of deserted or shrunken villages.

By 1700, therefore, we have a countryside and a farming system over much of England that we would recognize today. Much of the open field was enclosed, and with enclosure farms were moving out of the tightly packed village street into the fields. Farm houses and farm buildings were being modernized

37a *Hollowbank, Kentmere, Cumbria, showing the small farms scattered up the hillside. In the foreground is a stone byre at one end. Stone walls enclose patches of improved grazing (J.C. Barringer)*

37b *Bank barn, Satherwaite, Cumbria. Cattle are housed below the barn. It is entered from the far side at the upper level. The second floor window was once a winnowing door. The roof has been raised incorporating a row of ventilation slits (J.C. Barringer)*

38 *The Old House at Allendale, Northumberland, was built in the early seventeenth century in the bastel tradition, but of slightly superior status, as is shown by the mullioned window and moulded door head (Peter Ryder)*

or rebuilt in durable materials, and a variety of farm buildings symptomatic of the agricultural regions which are still with us today, were being erected.

Summary

The sixteenth and seventeenth centuries are the period when enclosure of the open fields of lowland Britain began, and these early fields can be distinguished by their mixed species hedgerows. Those fields resulting from piecemeal enclosure are likely to be long and sinuous, following the boundaries of the strips or furlongs. Increased prosperity of the yeoman farmers resulted in the building of substantial houses and farm buildings. As fields were enclosed, farmers moved out into their fields and a further wave of isolated farms were built. It is from this period that the earliest farm buildings, other than those of the monastic and other great estates, survive.

7 An agricultural revolution? The landscape evidence

Our study of the farming landscape has now reached that period of agricultural development which has long mesmerized historians. The period 1760–1830 was seen as an 'agricultural revolution', only comparable in importance to the 'industrial revolution', then seen to cover the same period. Since, the isolation of this period as one of major change, by Lord Ernle in the early years of this century, further research has thrown increasing doubt on this tight date range. Some would push the period of crucial change back into the sixteenth century, while others regard the period after 1830 as the more significant.

Written evidence is difficult to come by. Estate accounts were only kept by the best organized landlords, and similarly the publicists and travellers of the period only visited the great and famous. Inventory evidence (from the lists of possessions made when establishing probate and proving a will) certainly covers a greater cross-section of the population, but even this becomes less available through the eighteenth century. It may indeed be that that the only all-embracing source of information is the landscape, and there are some aspects of change associated with the 'agricultural revolution' which are obvious here.

First there is enclosure. This has already been discussed in its seventeenth century context and had been carried out since the Middle Ages. Gradually the open fields were being replaced by an irregular patchwork of boundaries interspersed often with a few blocks of strips, all of which was often reorganized in a more orderly fashion in the early nineteenth century and this gradual process of enclosure can sometimes be followed through estate maps. Much open field survived in 1800, and Arthur Young reminds us that it was part of European tradition. 'A village of farmers and labourers surrounding a church and environed by three or four and in a few cases, five open and extended arable fields form the spectacle of Cambridgeshire, Huntingdonshire and Northamptonshire as much as the Loire or the plains of Moscow'.[1]

The division between enclosed and unenclosed systems is not always as clear cut as it might at first seem. A lease for a farm in the north Norfolk parish of Burnham Westgate, in 1737, stating that not more than two corn crops could be grown in succession, was for 70 acres (28 hectares) of 'parcels of land lying dispersed in the several fields'. Another stated that at least half the olland (or temporary grass) was to be in the open fields, both suggesting that there need be no difference in farming practice between the open and closed fields.[2]

On the Holkham estate in Norfolk there were areas in the 1770s where the ownership of strips had survived, but the cultivation of the soil was not affected by their survival. An estate map of Kemptone, on the Holkham estate, shows areas of open strips,

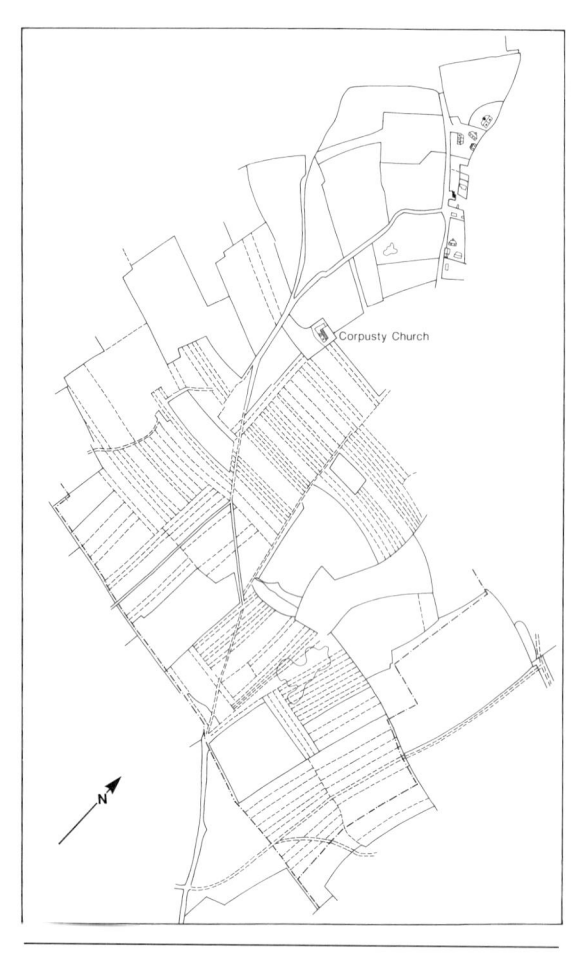

39 *A map of Corpusty, Norfolk, showing the strip fields surviving into the eighteenth century*

but a note in the accompanying survey book explains that the implications were purely legal. Some of the parish was owned by the landlord and some intermingled strips by the tenant. As a result, the area had been put into fields and the open-field baulks destroyed. However, it was still legally open-fields and so these boundaries were marked on the map. It is quite likely that maps may often show an archaic field structure for legal purposes, although in fact the land was farmed differently.[3]

This blurring of the division between open and enclosed field systems can make it difficult to assess how much open field still survived in reality, in 1760. As late as 1760, on the smaller Heydon estate, one farm contained strips, 'grazing grounds', 'new improved land' and 'half year land' over which others might have rights of grazing in the winter months, as well as some enclosed closes.[4]

From the mid-eighteenth century a method other than piecemeal enclosure or enclosure by agreement became commonly, but not universally, used. This involved obtaining an act of Parliament, for which, after 1766, the agreement of the owners of four-fifths of the land in a parish was required. This meant that a few large landowners could force their wishes on the smaller farmers, even though they might be in the majority. Most enclosures resulting from parliamentary acts covered commons as well as the open fields, but it is not often clear how much reorganization of the fields was involved. In some cases the act simply tidied up and finished a job already half done. Again, on the Holkham estate, Castle Acre and Tittleshall had no enclosure acts, whilst Fulmodeston did, yet Fulmodeston had no strips in the eighteenth century, and the other two did. It is incorrect to assume that a late eighteenth century or nineteen century enclosure act implies that little enclosing had taken place in the relevant area earlier on. Similarly, enclosure by agreement still continued. It was, after all, cheaper than obtaining an act of Parliament. These problems may be one of the reasons why Arthur Young regarded his estimate of the acreage of common field still surviving in the 1790s (4,600,000) (1,862,000 hectares) as being based on 'a very insufficient authority'.[5]

The division between wastes and commons and enclosed fields is far more clear cut. Between 1760 and 1800, nearly 600,000 acres (243,000 hectares) of waste in England and Wales was enclosed. Enclosure of common by agreement was far more

An agricultural revolution? The landscape evidence

difficult, and in many areas these were the only areas left for parliamentary enclosure acts to deal with. However, there were many landowners who took advantage of the enclosure of the common to rearrange all the fields in the parish on a more rational basis, and so there is no doubt that the landscape of a vast majority of estate-dominated parishes across England and Wales changed dramatically between 1760 and 1815. During this period eighteen per cent of the land surface of England was enclosed by Act of Parliamant. This mainly took place in two short bursts of activity, between 1760 and 1780, and between 1795 and 1815.

There were great regional variations in the pattern of enclosure. There was little parliamentary influence in the counties of the Welsh borders, the south-east and south-west, whilst it was in the first phase that much of the midland clay belt, as well as the lighter clays of Lincolnshire, and sixty per cent of the East Riding of Yorkshire, were enclosed. Between 1760 and 1815 more than fifty per cent of Oxfordshire,

40 *The parishes of Weasenham and Wellingham, Norfolk, were completely re-laid out by the Holkham estate in the early nineteenth century. The farmsteads were placed in the middle of their rectangular fields and roadways laid out so that every field could be reached from one. Where some fields were a distance from the farmstead a field barn was built (Phillip Judge)*

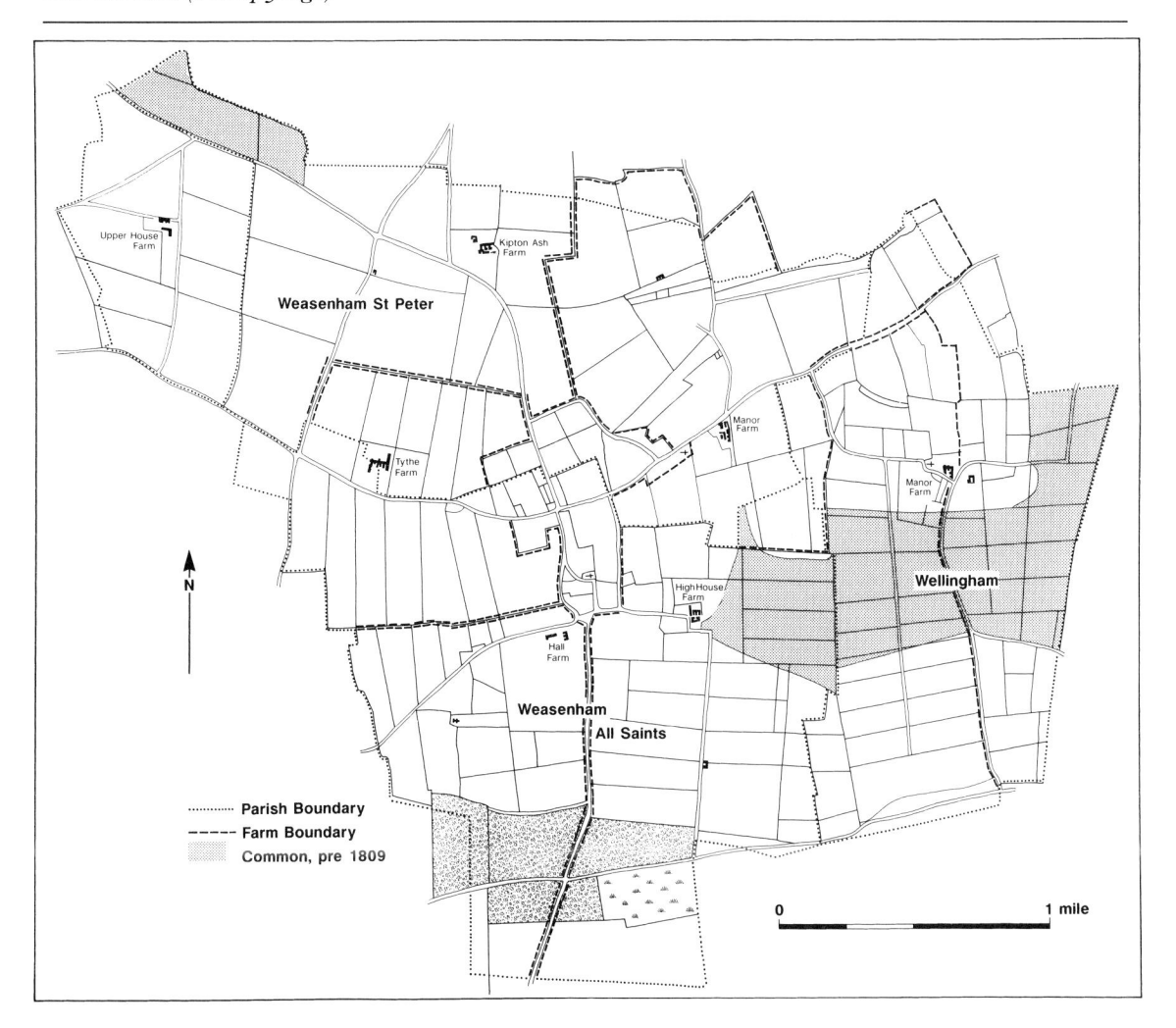

Cambridgeshire and Northamptonshire were enclosed as a result of parliamentary acts. Nationally, 200,000 miles (222,000 kilometres) of hedges were planted between 1750 and 1850 – as much as over the previous 500 years.[6]

However, it was not just soil types that influenced the chronology of enclosure, but also the landowning tradition of an area. 'The stronger the landowning tradition, the later the enclosure'. In Warwickshire, the pre-1750 enclosures were mainly by the squirarchy to consolidate their estates, 'whilst after 1750 the inspiration behind enclosure often came from freeholders trying to improve farming in general'.[7] In a period when the grain market was stagnant, as it was between 1760 and 1780, it is likely that these new fields were used for pastoral rather than arable farming, continuing the old tradition of the region which had led to unrest in Tudor times.

It was the second phase, between 1795 and 1815 (dates which coincide with the Napoleonic Wars), which was particularly impressive. Forty-three per cent of all acts date from this period, and not only the light soils of East Anglia, Lincolnshire and Yorkshire, but much marginal land in the Pennines, the Lake District and on the heaths of the southern counties, were affected.

The 'ancient enclosed grounds' of much of the northern uplands only occupied the dale bottoms and lowest slopes, not more than two or three fields in width from the river, ending as an irregular line along the dale side, known as the 'fell wall'. Beyond were occasional outlying islands of farmland. The moor was used for grazing, peat cutting, moor-coal working, quarries and lead mining, while long distance tracks used by packers and drovers crossed it. Much of this was to change with the improvements at the end of the eighteenth century. Firstly, the old vegetation was burnt off the moor to be enclosed. Then it was drained and the acid soils limed. Many farms of the moorland fringe were expanded or new ones created. Five new farms were created by Sir Edward Swinburne on his 14,000 acre (5,700 hectares) estate between the Scottish border and the North Tyne. The new stone-built and slated farm building included barns and granaries, suggesting that grain crops were hoped for in an area where there was over 50 inches (127 centimetres) of rain a year, and the land was between 600 and 1,000 feet (180 and 300 metres) above sea level.

In this period, when most of the commons and wastes were enclosed, 'the conquest of waste and the conquest of France became synonymous in some minds'. In 1800, perhaps twenty-one per cent of England and Wales was wasteland, whilst by 1873, this had been reduced to between six and seven per cent. It was the dramatic increase in grain prices during the war, from above £1 a quarter in 1790, to a peak of £6.50 a quarter in 1801, which encouraged this outburst of activity, and so these enclosures were for grain rather than livestock production.

All landowners wanted to share in the bonanza and there was no shortage of tenants prepared to pay high rents for farmland. Rents could double on newly enclosed land, and allowing for the cost of enclosure, which was probably in the region of £12 an acre, the return to investors could bring a fifteen to twenty per cent profit for the landowner.

Enclosure was certainly the aspect of 'the spirit of improvement' of the Agricultural Revolution which resulted in one of the most dramatic landscape changes of our agricultural history.

The principle of planned geometric layouts had been established before 1760, in those enclosures by agreement laid out by surveyors, and could be seen, for instance, in sixteenth and seventeenth century Leicestershire and Northamptonshire. The advantage of this degree of planning is explained in Stone's description of Lincolnshire agriculture in 1794. 'The first great benefit resulting from enclosure is contiguity, and the more square the allotments are made, and the more central the buildings are placed, the more advantages are derived to the proprietors in every respect'.[8]

Unlike the earlier enclosures, those of the late eighteenth century did not result in depopulation. They were usually for arable and so a large labour force was still needed. However, they did result in the creation of new farms in the new fields, sometimes called, Heath, Common or Allotment Farms, or after American War of Independence or Napoleonic War battles. The fields themselves were laid out with a geometric rigidity which swept away earlier boundaries. However, the commissioners were only responsible for the boundaries of each proprietor's allotment. Where there were only a few proprietors, there would be very little for them to do. It was up to the individual owners to arrange the fencing within their alloted holdings.

Fencing and hedging was an expensive task. 27,000 acres (11,000 hectares) of high Mendip was enclosed between 1770 and 1820, which probably involved about 1,650 miles (2,640 kilometres) of fencing. Much of this was in the form of dry-stone walling, which was cheap, especially where fields had to be cleared of boulders anyway. However, some farmers preferred the shelter that hedges provided for stock, and planted quick sets in the shelter of the wall. The walls would then be removed as the hedge became established.[9]

Sometimes the landowner took advantage of the enclosure of common to relay out the whole parish, whilst elsewhere there is a contrast between the straight roads and rectangular fields of the old common area, and the windey lanes and irregular layout of the rest of the parish. The old common edge can often be distinguished by the line of older houses set well back from the modern enclosure road. Along this road will be a few nineteenth century cottages or farms built following the changes.

The creation of roads 40–60 feet (12–18 metres) wide, was always an important and expensive part of any enclosure, and road building could amount to twenty-five per cent of the cost. This width was necessary so that there would be a reasonable way through the mud and pot holes, even in wet weather. As road techniques improved, roads could be narrower and by 1800, 45 feet (13.7 metres) was more usual.

An integral part of the enclosure was the improvement of the land. On the dry and heathy wastes of East Anglia this would involve 'marling', or the spreading of the chalky-clayey sub-soil on the surface, to put 'heart' into the tilth. About thirty loads to the acre (twelve loads to the hectare) was the recommended amount and this heavy, labour-intensive work was usually undertaken in the winter, often in wet conditions. The responsibility for this expensive and time-consuming work was often passed on to the tenant who could well find it a discouraging task. The parish of Longham, on the Holkham estate in Norfolk, was enclosed in 1816, just as the Napoleonic War had ended and prices began to fall. The old common was, as was often the case, the poorest land in the parish and needed marling and deep ploughing. When it was visited by the estate's agent, he wrote, 'Much money has been expended on the improvement of this farm, with a very far distant prospect of return . . . Mr Hastings is a zealous and industrious tenant, but heart-broken by his present undertaking'.[10] The remains of the marl pits, now frequently filled with scrub and a haven for wild fowl, remain in many an East Anglian field.

The burning of lime for the improvement

41 *An enclosure road across a heath enclosed in the early nineteenth century at North Elmham, Norfolk. The house on the right, set back from the road, is on the old common edge*

of acidic soils was important in the south-west and much of highland Scotland. The ruins of lime kilns remain along the Bristol Channel coast and along the River Torridge south of Bideford, Devon, burning limestone imported from the Gower Peninsular and Caldy Island, South Wales. Monumental lime kilns also survive through much of northern Scotland, such as on the Airde, Loch Shin, Sutherland.

In wet areas, improvement involved drainage. A Commissioners' drain runs across the Norfolk Broads and many new wind pumps were erected as a result of enclosure acts.

Although efforts had been made to drain the Somerset levels from Roman times, mainly by building sea defences, most of the work inland was piecemeal, and on a small scale. Up to 1770, 45,000 acres (18,000 hectares) of unimproved common grazing was often under water for half the year. This was all to change over the next fifty years with the passing of a series of enclosure acts, resulting in the digging of a 12-mile (19-kilometre) new channel for the river Cary, known as King's Sedgemoor drain. Although this helped bring new areas into cultivation and drain the pasture, it was not a total solution to the problem. Silting meant that the water could back up at the mouth of the river, leading to flooding.

The pattern of fields created was not as regimented as that usually associated with parliamentary enclosure. Because of the large number of people with rights to common grazing there were scores of tiny strips, resulting in an extremely complex pattern. New farms were built out in the drained peat, often of brick rather than the

An agricultural revolution? The landscape evidence

42a and b *The pumping station at Ten Mile Bank in the Fens. The plaques on the wall record each time a new engine was installed*

more traditional stone of the older buildings.[11]

The legal process of gaining an enclosure could be long, complicated and expensive. It cost about £3 an acre before 1760, but there was a considerable increase between 1760 and 1820. It could take up to six years, during which time a farmer did not know what was happening or how to farm. After about 1760, the interpretation of the law made it easier for an agreement to be reached. The agreement of those who held the majority of the land, rather than the majority of landowners, was needed, meaning that the few large landowners could over-rule the many small owners. A public notice of intent to enclose was placed in local papers and then the commissioners were appointed. These men, often JPs from neighbouring villages, attempted to fulfil their task of establishing claims to land fairly, but the lack of written title deeds meant that many smallholders lost out, as did squatters on commons who had no status in the eyes of the law. Arthur Young described the commissioner as 'a sort of despotic monarch, into whose hands the property of a parish is invested to recast and distribute it at his pleasure among the proprietors and in many cases without appeal'.[12] The land was then divided up by surveyors, and the individual owners given the task of planting hedges within and around their plot. The cost of hedging a smallholding was disproportionately high as compared to a larger piece of land. For these, and other reasons, based on the economies of scale of the improved farming, small owners were bound to lose out and many sold their allotments to the larger farmers, thus speeding up a process which was probably inevitable. It would be wrong, however, to assume that the small owners were necessarily occupying and farming their land. Many were outsiders, perhaps local tradesmen, who owned and let small acreages, often scattered around several parishes.

The national investment in enclosure amounted to about 29 million pounds, more than that in canals (20 million), but far less than the national debt resulting from the French wars (500 million).

The debate on the social and economic consequences of enclosure has been a long one, dating back to the time of the enclosures themselves. Arthur Young was convinced that enclosure should result in improved agriculture. Of Knapwell in Cambridgeshire, he wrote, 'the improvement is not supposed to have been so great as it ought to have been, and no wonder as they cultivate the land nearly as before'.[13] There was certainly an increase in arable farming, particularly during the Napoleonic Wars, as commons and waste were brought into cultivation. The sheep walks of the Yorkshire wolds, for instance, were enclosed for arable farming. Attempts to grow crops on the newly enclosed light heathland of Breckland were short lived and mostly abandoned after the restoration of peace in 1815, but not before huge barns had been built in anticipation of good crops.

They stand as a monument to the optimism of the owners.

The new rotations meant that the individual crops of wheat, barley and oats were more likely to be grown one in four or five years rather than the one in three in the common fields, and so here there would have been a fall in wheat, oats and barley acreages. However, increases in yields would have ensured there was no fall in output. Pulses on the other hand declined in favour of the improved grasses. Enclosure encouraged the efficient management of the land, but its importance should not be over-emphasized. Certainly turnips and other forms of new husbandry had penetrated the open fields by 1720, but their efficiency must have been much reduced in the dispersed holdings of the system. There was also an increase in the quality and quantity of livestock kept, which in turn allowed for more manuring of the land. There must have been considerable improvements in output otherwise the farmers would not have been prepared to pay the higher rents demanded, sometimes treble those on the open-fields. Although the farmers had to learn new skills, particularly in the management of livestock and pasture, the long term benefits were worth it. As Arthur Young put it, the open-field farmer was 'in chains', 'a mere horse in a team – he must jog with the rest'.[14] We have already seen that this colourful metaphor is something of an exaggeration, but no doubt there is some truth in it.

That the effect of enclosure on the poor could be serious is shown by the fact that the poor rates were generally lower in areas where the poor still had the right to graze a cow on a common. There was an increase in landless labourers relying on wages for a living, but there should have been work for them on the new labour-intensive farms. In the words of Michael Turner, 'Enclosure certainly shook the countryside up, but it was population increase which was really responsible for social dislocation and the move of population to the towns. Enclosure lubricated the process, but was not its sole, nor necessarily its main cause.[15]

Following on enclosure, the next most striking landscape change of the period was the provision of new farm buildings. It would be wrong to assume that new building necessarily followed an enclosure act. Regional studies have shown that farms unaltered by enclosure were just as likely to receive new buildings as those whose fields and acreages were re-organized.[16] However, where new farms were created or where scattered holdings of strips were swept away, leaving all the farmsteads in the village, then of course new buildings were needed. William Pitt described, in his account of the agriculture of Northamptonshire, open-field villages where the farmsteads were 'pent-up in villages and are consequently either on one side of the farm or totally detached from it.' In the Mendips new farms were not built until the 1820s or later, although the enclosure was somewhat earlier. Initially the newly enclosed land had been added to the already existing lowland farms that had had grazing rights on the old commons. However, this was unsatisfactory, and finally landowners created new farms with houses built of the local grey stone with tiles or slated roofs, flanked by barns and shelter sheds on one side and a waggon lodge with granary above on the other, sometimes with a pig or poultry house as well. It was calculated that to build a house and buildings for a 400–500 acre Mendip farm cost £1,300.[17]

There was much criticism of the state of farm buildings at the end of the eighteenth century, and particularly of their lack of planned layout. They were scattered, 'at random ... without order or method, whose buildings had accumulated over generations'.[18] This is very likely to have been the situation on owner-occupier farms where the capital for a complete rebuild was unlikely to be available, or even a tenanted farm where the role of the landlord as provider of fixed capital was not recognized. By the late eighteenth century, and particularly, as we have seen, as the result of enclosure, the role of the landlord was becoming more clear-cut. The tenant, often holding his farm at fixed rent on a long and secure lease, was responsible for his working capital for tilling the soil and keeping enough livestock to ensure the land was well manured. The landlord on the other hand, provided the buildings, fences and ditches for the tenant to keep in repair.

The cost of buildings was never as great as for the initial enclosure. Half the expense of enclosure could go on hedging and fencing and only a sixth on farm buildings.

As profits and rents rose, the landlord had money to re-invest and he was well aware that if he wanted to attract the best tenants, he needed to provide good buildings and particularly, good houses. Fine, classical houses with slate or tiled roofs were built. The principal rooms overlooked a park-like home pasture with the garden often protected by a ha-ha, while the back of the house looked straight onto the cattle yards. The tenant might well imagine himself a gentleman when standing in the drawing room of his house, but he was definitely a farmer when he went to the back.

New farming techniques too meant that new buildings might well be needed. Increased acreages and yields meant a greater demand for barn space in which to store part of the crop, and to thresh out the grain with flails throughout the long winter months. By far the majority of surviving barns in England, from those serving the smallest to the largest farms, are eighteenth century. Many of these are simple structures, built of local materials, their only distinctive feature being their opposing, usually double, doors, allowing a laden waggon to enter, and for the creation of a good draft to winnow the chaff away from the grain. Their plain

84   *An agricultural revolution? The landscape evidence*

43 *Thorndon Hall, near Brentwood, Essex, designed by the country house architect, Samuel Wyatt in 1777*

44 *Another example of the work of Samuel Wyatt, this time on the Holkham estate at South Creake, Norfolk, built in the 1790s (Phillip Judge)*

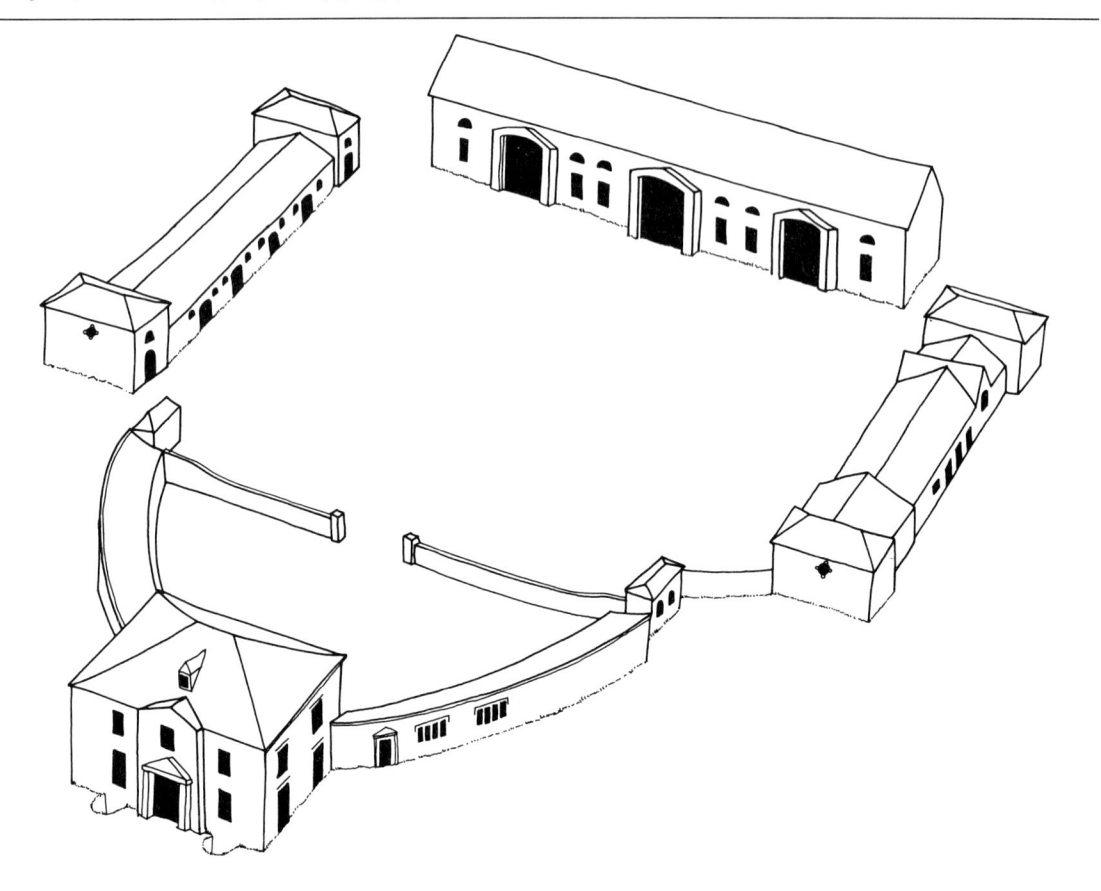

An agricultural revolution? The landscape evidence

windowless walls are unbroken except for ventilation slits, and their interiors are huge and open. Their roofs are of straightforward construction, using wooden pegs and morticed joints, illustrating the last phase of vernacular carpentry before the introduction of soft-woods, roof trusses, iron nails and tension rods.

Increased yields were achieved by the keeping of more animals and whilst there was little housing for sheep, cattle were housed, either in open strawed yards and shelter sheds, or in the upland and western parts of the county, indoors in loose boxes or tied up in byres. Whilst the tying up of cattle in byres had gone on from the earliest times, the building of open sheds and yards was an innovation of the eighteenth century. The earliest sheds identified by Dr Peters in Staffordshire were dated 1754.[19]

The best layout for this group of buildings – barn, cattle sheds and open yard, often with stables and granaries on one side, was around a courtyard with the barn on the north side, providing shelter for south facing yards. Many such groups of buildings were erected from very simple examples to some architect-designed ones, built with the

45 *Late eighteenth or early nineteenth century farmyard at Home Farm, Roundway, Wiltshire. The barn and house are on the left with an earlier dovecot with stables below, flanked by cowsheds and shelter sheds to the rear of the yard. It is unusual for a farmyard to remain open as this one has, uncluttered by later divisions. It was largely demolished in 1980 (Wiltshire Buildings Record)*

46 *The home farm was often the show-piece of an estate and that at Holkham, Norfolk, was no exception. Again, it is the work of Samuel Wyatt in the 1790s. The extent of the granaries is some indication of the importance of the cereal crop*

intention of impressing all comers with the status and wealth of their builders and owners.

Such courtyard layouts were usually confined to farms over 150 acres (60 hectares) and so are more likely to be found on estate or gentry farms than on small owner-occupier ones. They were by no means universal in the period before 1840. In selected survey areas of Norfolk only eleven per cent of layouts could be classified as U-shaped and half were irregular.[20] Layouts of the smaller upland farms were more likely to remain linear with the highly practical bank barn being developed in the eighteenth and nineteenth centuries. This compact building, combining both cereal and livestock accommodation, is surely one of the most functional of farm buildings, and is found throughout the upland western regions of Britain, from Devon to the Hebrides, with a concentration in the Lake District.[21]

The development of housing for cattle was swiftly followed by other technical advances affecting farm layout. Turnip- and straw-chopping machinery was being developed from 1760, and many farm plans included turnip houses and other fodder rooms from 1770. The boiling of feed was being recommended by the 1790s and food preparation rooms appear on some plans. From the 1780s, threshing machines were available which could be powered by horse, wind or water power. In the 1790s, the use of steam was being pioneered. However, the adoption of mechanization was much slower than the acceptance of the courtyard plan. The 'agricultural revolution' of the late eighteenth century was mainly concerned with an improvement of husbandry techniques stimulated by enclosure, rather than mechanical innovation.

In some areas the progress of mechanization was faster than in others. In

An agricultural revolution? The landscape evidence

Scotland, Durham and Northumberland, where there was a shortage of labour, particularly near the expanding mines, threshing machines were adapted quickly. When William Cobbett visited Berwickshire in 1832 he found that all threshing was done by machine, 'there being no such thing as a barn or flail in the whole county.'

The important and unique role in Europe that the British landed estate and its owner took in the promotion of improved agriculture in the late eighteenth century is best reflected in the 'model' farms. The work of the landlord and his architect in their creation is 'perhaps the most surviving monument to this particular civilization and its ideals' and has been fully described by John Martin Robinson in his book, *Georgian Model Farms* (1983). He could only find about 300 examples in England and Wales, and fewer in Scotland, to list in his gazetteer. Their distribution across the country is not even. They are mainly to be found in areas of large estates and where arable farming was particularly profitable, concentrated in Norfolk, Lincolnshire, Yorkshire, Northumberland and East Lothian. They are mostly confined to home farms within a park or near the great house, and served as a showpiece, not only for the tenantry, but also as a place to be visited by the owner's guests. On some estates, however, notably those of Sir Christopher Sykes at Sledmere, the Marquis of Stafford in Staffordshire, Shropshire and Sutherland, The Duke of Bedford in Bedfordshire and

47 *The Home Farm at Wimpole, Cambridgeshire, was designed by Sir John Soane in 1794 and is an early example of a planned farm (redrawn by Phillip Judge)*

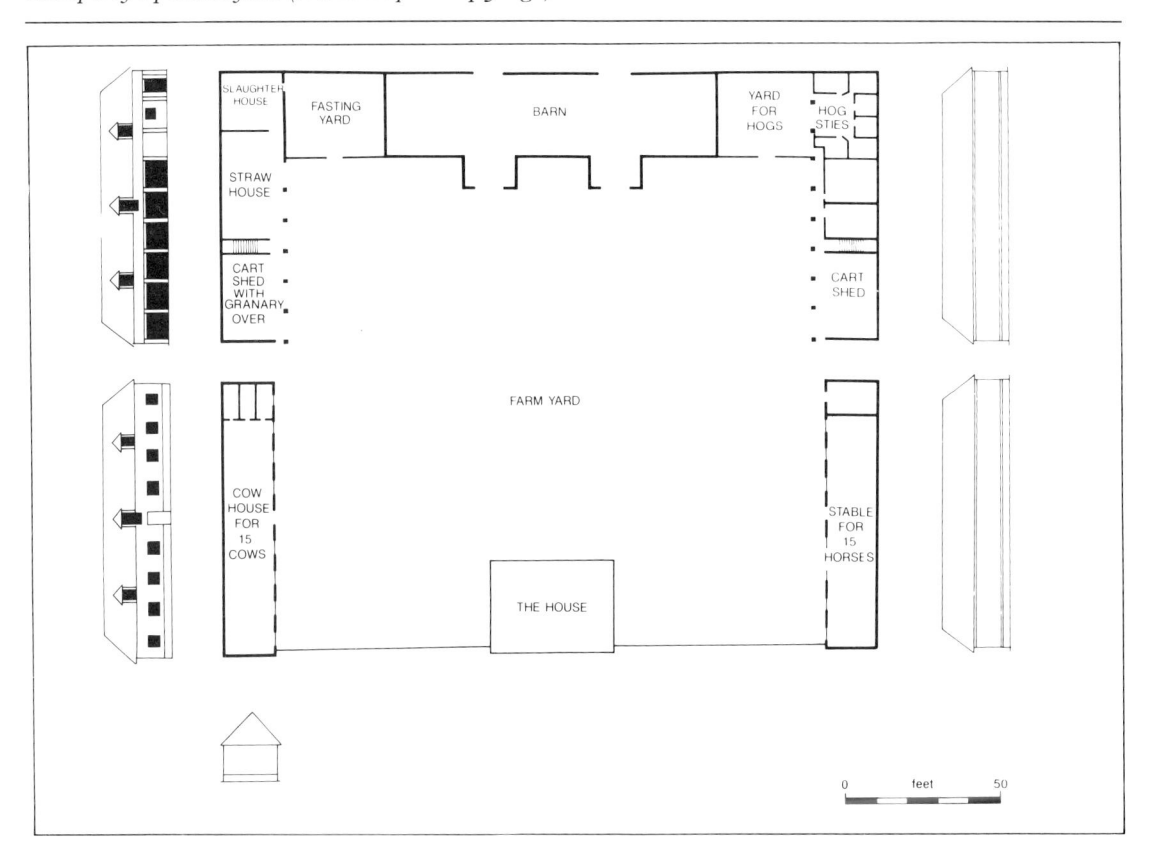

48 At Chollerton farm, near Hexham, Northumberland, is one of the few remaining windmills installed to work a threshing machine. That this was not reliable enough is shown by the fact that a steam engine was later installed

Cambridgeshire and Thomas William Coke in Norfolk, some of the tenant farms were given similar treatment.

Courtyard plans with a large barn, shelter sheds, stables, cart lodges and feed stores were erected. In the earliest examples, gothic facades were often created, as at Rousham in Oxfordshire where a castellated cattle shed is part of the formal park, and at Badminton, Gloucestershire, where two farms and one barn form 'the best group of mid-eighteenth century rococo Gothic farm buildings to survive. Workshop Manor Farm in Nottinghamshire was designed by the Duchess of Norfolk in about 1760. Its huge courtyard, 66 yards square (60 metres square) was surrounded by castellated barns and cowsheds, making it an eye-catcher in the park and visible from the house.

By the end of the eighteenth century, Gothic was giving way to neo-classical and this resulted in some of the finest farm buildings in Britain. Many owners employed architects and amongst these, Samuel Wyatt, who worked at Holkham and Shugborough as well as for many other smaller estates, was one of the most prolific. His Great Barn, built at Holkham as the backdrop for the sheep shearings is one of his most famous buildings. It is in a severe classical style of yellow brick with a slate roof.[22] The huge barn with its two impressive porches still stands, but the cattle sheds that surrounded it, no longer survive. Here many of the events of the annual sheep shearing held from the 1770s to the 1820s, took place. Stock was inspected and sheep shorn in front of an audience that grew over the years, from being merely local tenants to include the gentry and aristocracy of Britain and Europe.[23] Other landlords, such as Sir Christopher Sykes, designed their own buildings,[24] often relying for their inspiration on the designs produced in the increasing number of pattern books of which those of J.C. Louden were amongst the most popular. The advice given was often contradictory, from an intensely utilitarian view stating that 'farm buildings . . . should be simple in their form and perfectly plain', any appearance of elegance being created by their 'regular arrangement and modest neatness', to an extremely decorative approach.

It was the landowner too who was able to indulge in flights of fancy as far as ingenious

An agricultural revolution? The landscape evidence

forms of mechanization were concerned. Whilst this was particularly true later on, some landowners were incorporating water power into their farmsteads by the early nineteenth century. White Barn Farm at Shugborough was extended to form a symmetrical group of buildings between 1800 and 1806 and contained the first threshing machine worked by water power in Staffordshire.

On the whole, however, developments were confined to the large farms owned by progressive landowners. The smaller yeomen would have adapted to the changing times by adding a bay to his barns, rebuilding the stables and putting up some open-shelter sheds, providing cover as best he could. His concession to agricultural progress would be the growing of some turnips and improved grasses in rotation, thus ending the fallow year. This more intensive farming would necessitate the keeping of extra horses to work the farm; the production of more grain meant more storage space was needed; and the keeping of a few more cattle provided the manure to increase production. The only evidence for the type of farming practised and the acceptance of progress by these men is to be found in their buildings. Many could not write and few kept diaries. The only written evidence may be in the details of the sales of their live- and dead-stock when they left their farms. Notices of such sales were placed in local papers and their goods often consisted of no more than 'four useful carthorses, a mare and foal, six profitable cows and ten store pigs' as well as a number

49 *Sir Christopher Sykes was responsible for the design of his new farms in the Yorkshire Wolds. The illustration shows Maramette Farm at Sledmere built in the 1770s (Country Life)*

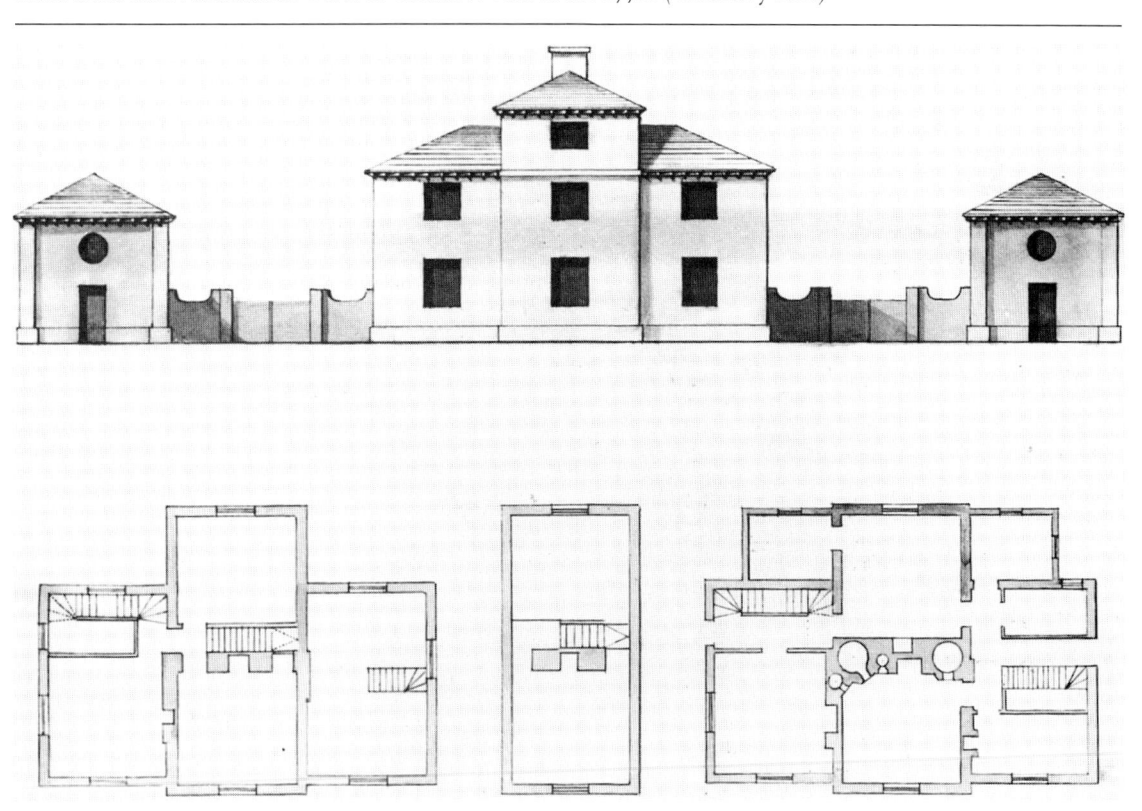

of fowl. The equipment included a waggon and cart, a plough, harness, a land roll and dressing machine as well as dairy and brewing utensils. It is these smallholders who would have made up the majority of farmers in the United Kingdom, and their small-scale buildings characterize the farming landscape.

To return to the question posed at the beginning of this chapter, there certainly was a 'revolution' in the appearance of the landscape over much of lowland Britain. Even where there was no enclosure act, fields were rationalized and re-arranged. A new landscape, which has survived to the present, was created. We have already seen that the complete wiping away of the old was not as total as has sometimes been imagined, with prehistoric and Roman landscapes underlying later changes. Certainly, however, the early nineteenth century was the last period of major landscape change before the present.

The situation with buildings is rather different. Although there are examples of splendid model farms that have survived almost unaltered to the present day, these are the exceptions. This early phase of agricultural change involved developments in husbandry rather than mechanization and it is the later, second phase that resulted in far more important changes and developments in farm building design.

Summary

This is the period of the landlord's agricultural revolution and his work of 'improvement' often transformed the landscape of his estates. Enclosure fields with their thin hawthorn hedges surrounding rectangular fields, and straight enclosure roads dominate. New farms can be recognized by their plain Georgian houses and barns set up along a straight drive off the road, in the centre of their fields. Some early nineteenth century cattle yards survive as part of the original planned farms, but most were replaced later in the century. Mechanization in the form of horse works, windmills and steam engines also found its way on to some farms.

The spirit of improvement not only affects the great estates, but everywhere farms were rebuilt and an estate 'style' was adopted. The insignia of the owner or a distinctive type of brick or cast-iron work can often be identified over a limited area of countryside.

Away from the estates changes were taking place, but they were more likely to be slower and less dramatic. More older buildings are likely to survive in the owner-occupier than estate dominated regions.

8 The Welsh and Scottish experience

Wales

The progress of the 'agricultural revolution' in Scotland and Wales was rather different from that in England.

Open-field agriculture and its associated nucleated villages were never very important in Wales, being confined to the valleys penetrated by the Normans and the easily defended lands near castles. As the lords' demesne was being let and the old bondsmen became tenants there was much piecemeal enclosure as independent farmsteads were created. In the predominantly Welsh areas, holdings were small and scattered, and here consolidation began in the fourteenth and fifteenth centuries. The vale of Glamorgan was enclosed in the seventeenth century and scattered farms of between 25 and 70 acres (10 and 28 hectares) created as the landscape took on its present appearance. Many farms, especially those where there was unlimited mountain grazing, were far smaller, and 15 acres (6 hectares) was regarded as enough to support a family. Whilst cattle were the mainstay of lowland agriculture, sheep were becoming dominant in the mountains where oats was the only cereal grown.[1]

The survival of Welsh 'law books' helps in the reconstruction of the farming pattern across the country, suggesting a communal system with shared-plough teams, scattered strips divided from each other by baulks about 18 in. wide (46 cm), and common grazing. The ideal size hamlet in the fifteenth century, according to these books, consisted of nine houses, one plough, one (grain) kiln, one church, one cat, one cock, one bull and one herdsman.[2]

Highland over 600 feet (180 metres) played an essential part in the farming system as the cattle and sheep would be driven up to summer pastures, known as sheilings, and summer houses or 'hafods' would be occupied. Traditionally the livestock were kept in the uplands from May to October and then returned to the 'hendre', or winter dwelling. The system is described in Snowdonia in 1770, but was generally in decline by the eighteenth century. Some hafod sites were deserted whilst others became permanent settlements taking in even higher land for summer grazing. The distribution of names with 'hafod' and 'hendre' is sometimes the only evidence for this medieval system. The permanent settlement of hafods, and the piecemeal fencing of land around them, was one form of enclosure in the fifteenth and sixteenth centuries. Elsewhere, there was a nibbling at the edge of the great commons, both by individual farmers and squatters erecting a cottage on the common overnight, and so claiming a right to the land immediately around it.[3]

Woodland, moorland and open pasture was being enclosed with, for example, 2,000 acres (810 hectares) in Cyfeilig (Montgomeryshire), being enclosed between 1561 and 1573.

By the eighteenth century, certain changes were taking place. Farms were being amalgamated, and a social elite was emerging as larger estates were being accumulated. These estates were created, not only by the buying out of smaller neighbours, but as a result of intermarriage and failure of male heirs. In this way, Robert Watkins Wynne of Wynnstay, Denbighshire was heir to no fewer than five estates in 1792.[4]

A third of the total acreage of Wales was still common in 1795, and as elsewhere, it was the enclosure of waste that was important during the Napoleonic wars, bringing to an end the already declining transhumance system. Between 1793 and 1815, over 200,000 acres (80,000 hectares) of Welsh common, both wet lowland and high moor was enclosed, including 40,000 acres (16,000 hectares) of Brecon forest, but this did not lead to the creation of new farms. Rather, land was added to those already existing. The general poverty of these farms meant there was little new building before 1800, but as the great estates grew, particularly in the south of the Principality, there was much improvement up to 1850. In Pembrokeshire where ninety per cent of farms were estate farms, there was much rebuilding during this period. Few ornate model farms survive from the period before 1850, but an unusual example is the cruciform building at Model Farm, Wolvesnewton, on the Duke of Beaufort's estate in Monmouthshire. At the centre of the cross was a feed store and the wings probably provided cattle accommodation, served by a railway.[5]

As late as 1870 arable farming remained important with as much as forty to fifty per cent of the enclosed land of Wales under tillage in the 1840s and 50s. In Merionethshire (a prime pastoral county) it was the size of the rickyard rather than the number of cattle that determined the nineteenth century farmer's social standing.

Cattle were, however, the basis of the farming economy, and after 1870 there was a wide spread conversion to dairy farms. The byre was the most important building and there was a close association of byre size and farm size. Barns, on the other hand, were small and insignificant. The demand for labour in the mines of south Wales by the nineteenth century meant that the population of north Wales began to decline after 1841. The general shortage of workers for the land meant there was a need for mechanization. Water and horse wheels were installed on many farms. Farms in the more remote areas were less likely to be improved, and examples of more primitive buildings survive in Radnorshire, for instance. Gradually land, even in these areas, was consolidated into great estates and with these trends went the amalgamation of farms. Between 1860 and 1890 the number of farms halved. Many Welsh farm buildings date from this later period.[6]

Very little in the way of farm buildings survives in west Wales from before the eighteenth and nineteenth centuries, whilst in the south there are some seventeenth century farm houses. It is in the south that the oldest building types and indeed archaic farm practices such as ploughing with oxen, survived longest. The circular, corbelled pigsties to be found in the region are surely a prehistoric building type. About fifty longhouses also survive from the seventeenth century, although many were remodelled in the eighteenth. The example from Cilewent, Radnorshire (Powys), was rebuilt at the Welsh Folk Museum in the 1950s. Originally timber-framed, the walls were replaced by stone in the 1730s. The eighteenth-century house contained one living room/kitchen/workroom containing a large hearth in the gable wall. There was a locally-held belief that cows gave more milk if they could see the flames of a fire, and this was certainly possible in many longhouses if

50 *Longhouse at Cilewent, Powys (Welsh Folk Museum)*

the door to the living room was left open. The cowstalls were separated from the living quarters by a corridor with a half-height wall on the byre side and a wall and door on the house side. At Cilewent twelve cattle were stalled across the passage at the lower end of the house. Furthest away were the stalls for two horses. Above the stable and byre was a loft for storing hay and straw. Oats for both animals and humans was kept in a large chest in the passage way dividing the animals from the humans.[7]

Scotland

Changes in Scotland during the 'agricultural revolution' were far more dramatic than elsewhere, and the enclosure of a previously almost completely open landscape took place mainly in the fifty years after 1745. However, previous to this, in the seventeenth century, a certain amount of limited enclosure had been carried out, particularly in the immediate vicinity of the laird's house. By 1800 there was no legal barrier to enclosure, which could be carried out without an act of parliament.

Previous to that date the county was mostly farmed according to a run-rig system of cultivated strips administered by the 'fermtoun'. The fermtoun consisted of the houses and land of a group of people who cultivated the area communally using a plough team pulling the unwieldy old, wooden, Scottish plough. The land was divided into the 'inbye' or the land nearest the settlement, and the 'outbye' or more distant, poorer area, about four times the size of the inbye but often only a tenth of its value. It was only cultivated occasionally and instead was used for grazing or peat cutting. The 'inbye' was divided into long curving strips which were kept permanently cultivated and heavily manured from the cowhouses and household waste, and if near the coast, seaweed. There were no fences; instead baulks separated the rigs, which could vary from 220 yards to 445 yards (200 to 400 metres) long and 11 to 28 yards (10 to 25 metres) wide. No farmer held adjacent strips and the families lived in an irregular cluster of houses beside a small 'garden' usually enclosed by a stone or turf wall which was hand-cultivated. The inbye and outbye were also divided by a stone wall.[8]

Although there is little evidence for this early system of landholding in the modern landscape, the line of the dividing walls and baulks, particularly between the 'inbye' and 'outbye' often survives. When, for instance, the early maps of Craigton, a small farm near Stirling, are compared with the modern field pattern, it is striking how many of these early boundaries are respected today.[9]

This early landscape of small islands of cultivated land surrounded by a sea of rough untamed country, from which nearly all the trees had been removed, was typical of early eighteenth century lowland Scotland. One hundred years later nearly all the obvious evidence for it had gone, although, as can be seen in the case of Craigton, there may be more surviving than has always been realized.

The changes of the late eighteenth century had many causes, political, economic and technological. The old run-rig communal system had been developed for mutual protection in a period of unrest. It satisfied two 'elemental needs, self-defence in a lawless land and bare subsistence agriculture.'[10]

Some seventeenth century peel houses, reminiscent of the bastel houses across the border in Northumberland, with byres below and accommodation above survive in

51 *Pants' map of Craigyloach, Perthshire, showing the run rig and township before improvemnt (R.J. Brien)*

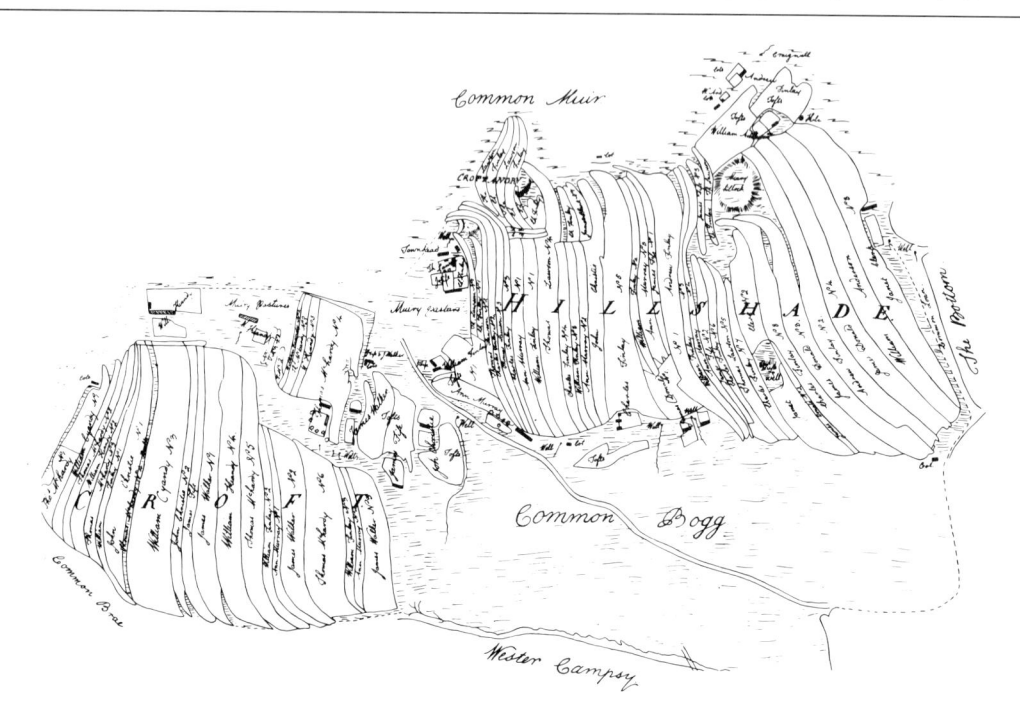

the border areas, particularly in the Jed valley, but they were becoming obselete as life was becoming more peaceful in the years after the Act of Union (1707). Possibilities for trade were also increasing as the independent burgs were being founded and cattle dealers, maltsters, clothiers, and skinners were becoming established in them. Gradually, as the conservatism of the lairds was broken down, the ways of the south filtered north and trade increased, so it was worth producing surpluses for sale. All this made the old fermtoun obsolete. The run-rig was divided into farms which were tenanted. This was followed by a wave of systematic enclosing, draining, liming, planting and building which totally transformed the lowland landscape.[11]

Before 1700 the granting of written leases, the payment of rents in money and the application of lime to the acid soils, so typical of much of Scotland, had begun, but only to a limited extent. Lime was being applied to some Lothian and Ayrshire farms by 1627, resulting in some instances in a five-fold increase in rent. Two societies for improvement were founded in the 1720s, both with great landowners amongst their members. The Honourable Society of Improvers in the Knowledge of Agriculture was founded in 1723, and The Board of Trustees for Fisheries and Manufacture in Scotland, in 1727. The work of these societies can be linked to the first generation of improvers in the years up to the 1745 rebellion. A limited number of landlords such as Lord Belhaven, the sixth Earl of Haddington and John Cockburn of Ormiston, all in East Lothian, began improvements, but financial difficulties forced Cockburn and some others like him to sell their estates. The second Duke of Argyll and his successors had more money behind them and by 1740 single tenancy farms were being established and new ideas adopted on their extensive estates on Mull, and from Kintyre in the south to Morvern and Tyree in the north.[12]

The real impetus for change came after the crushing of the last Jacobite rebellion in 1745, when the earlier ideas were taken up by more business-like men. Jacobite lairds were exiled and the lands of the rebel clans forfeited. The immediate consequences of the Hanoverian victory were dramatic, but of much more long term significance to the farming landscape and rural life was the work of the 'improvers' which followed. This was spearheaded by the Board of Agriculture, led by Lord Kames of Blair Drummond, which was set up alongside The Board of Management of Annexed Estates in 1752, and the Trustees for Fisheries, Manufactures and Improvement.

The annexed estates consisted of thirteen of the largest forfeited estates, which were singled out for special attention with the aim of eliminating the old social order. The Commissioners for the estates included twenty of Scotland's leading proprietors and their work stimulated others to follow suit. Lt. Col. David Watson produced for the Board of Management 'Instructions to Surveyors' working on the estates, and these came to be used widely by improvers throughout Scotland. The estates covered a huge area from Wester Ross to Perthshire, and here farms were re-laid out in single tenancies, trees planted, fisheries developed and linen and other industries introduced, churches and schools built and communications improved. These bodies lasted until 1784 when the forfeited estates were restored to their original owners.

The work of the commissioners contributed to the tranformation of agriculture. They were given a free hand to experiment on other people's property. On a wider scale, turnips were first introduced as a field crop in the 1740s and potatoes soon after. Wheat began to be grown further north, barley replaced the more primitive bere and better strains of oats and rye were introduced. Black-faced sheep penetrated further and further north and by the 1780s the Aberdeen Angus and Ayrshires were becoming established cattle breeds. More importantly, as far as the landscape was concerned, fermtouns were broken up and replaced by individual farms. Land was drained, marled, limed and cleared of boulders, and trees were planted. New farms were laid out and buildings erected. In parts of the lowlands the value of farms increased six times between 1769 and 1800.

One of the most valuable sources for studying the farming landscape of mainland Scotland is Roy's Military Survey made in the aftermath of the 1745 rebellion between 1747 and 1755. It shows with great accuracy the farms and cultivated area at the time. The height and remoteness of some of the farmed area, often over 1,200 feet (365 metres), is surprising and some of this was abandoned in the years up to 1800, and not reclaimed, even during the Napoleonic wars. On the other hand, there were large areas of wasteland and lower moors which remained uncultivated in 1750, but were brought into use theafter. 495,000 acres (200,000 hectares) of low-lying moorland was reclaimed for the first time. As a result, the farmed area of Scotland increased by about forty per cent between 1750 and 1825, whilst productivity rose by a hundred per cent.

In spite of this increase in productivity, fewer people were needed on the new individual farms. The new farmer, usually the strongest member of the original fermtoun community, ousted the weaker members who moved into the industrial towns or perhaps emigrated. The scale of this change can be illustrated in Kilmarnock parish where in the twenty-five years between 1765 and 1790, the population more than halved from just over 800 to 400. The most infamous example of such clearances were those in Sutherland, which will be discussed in detail in a later chapter. In some areas, the population was moved to planned villages where they were employed typically in either fishing or textile manufacture.

The new farms were held on fixed leases which stipulated that the tenant should enclose the land, usually with walls (although in parts of Perthshire ditches and quickset thorn hedges were more usual), and keep the buildings in repair. The onus for improvement was thus shifted through the lease, to the tenant, and by 1760, the large capitalist tenant had emerged and was responsible for much of the later stage of improvement.[13]

Once the movement towards improvement began it became unstoppable. Pamphlets were published from the early eighteenth century and agricultural societies were being set up in the late eighteenth century. In 1790 a chair of agriculture was founded at Edinburgh University; the first of its kind in the United Kingdom.

Most importantly, there was a rash of farm building between 1780 and 1820. The new farms were usually more extensive than their English counterparts and thus justified larger and more substantial buildings. Most of the buildings within the fermtoun had consisted of single-storey byre-houses built of mud or stone and roofed with peat or heather. The animals were kept at one end and the family lived at the other. These were systematically replaced by pattern-book stone farms built around a courtyard. In earlier examples the house formed one side of the yard, but later the house was more likely to be separate. In the more fertile areas the house was two-storied and substantial while elsewhere it was single storey, but always solid and well built. The entrance to the yard was often through an arch, above which was a dovecot. Some of these, particularly on the home farm or 'mains', could be very elaborate. Around the yard were the cattle sheds, and sometimes the labourers' single-storey cottages formed one side.[14]

The number of these sets of buildings which survive, particularly in East Lothian

52 *Semi-circular early nineteenth-century model cattle steadings at Aden, Grampian, now part of the North-East Scotland Agricultural Heritage Centre (Banff and Buchan District Council)*

where the money of many a wealthy Edinburgh merchant or banker went into estate development, is striking, even today. The existence of towns, fairs and ports as well as the large urban market of Edinburgh and a good transport system all helped to make this area one where improvement was on a scale comparable to the most publicized developments south of the border.

Although examples of late eighteenth and early nineteenth century farm buildings, sometimes plain and sometimes with classical facades, do survive throughout Scotland, many were altered and rebuilt in the mid-nineteenth century as mechanization and an increasing demand for cattle from the southern markets made the earlier buildings obselete. Some of the finest examples of the industrial buildings of 'high farming' are to be found in Scotland.

The changes in the north-west highlands and the Hebrides were different to those elsewhere and resulted, not only in the creation of single-tenancy farms, but of part-time smallholdings with both individual fields and common grazing, known as crofts.

The population of these areas increased by about a third between 1750 and 1800 and so there was a surplus of labour waiting to be utilized by the landowners. It was about this time that another source of wealth for the landlords of the coastal estates became apparent. This was the harvesting and drying of kelp to produce soda ash, valuable in the manufacture of soap, glass and particularly gunpowder. The price of kelp rose dramatically during the Napoleonic wars. Between 1808 and 1810 Reginald Clanranald made the enormous sum of £42,000 from the kelp produced just on his Hebridean islands of Eigg and Canna. Between 15,000 and 20,000 tons of kelp was exported annually from the Hebrides as a whole at this time. The landowners, anxious to make as much profit from the kelp as possible, needed a cheap, seasonal labour force, and the solution was seen as the creation of crofts. The population was moved from the inland areas which could now be let, often to farmers from the south, as large single-tenancy sheep farms, to the coast, where the new crofting settlements were laid out.

This type of reorganization probably originated on the Argyll estates, but spread rapidly, particularly during the kelp boom years of the Napoleonic Wars. The

53 *In the colder and wetter parts of Scotland, grain-drying kilns were still in use in the nineteenth century. This one is at Sibster in Wick parish, Caithness (Historic Scotland)*

54 *The same large farm also had a wheel house for a horse gin, later superseded by a steam engine (Historic Scotland)*

Clanranalds commissioned surveys of many of their estates drawn by the surveyor and engineer, William Bald. His large-scale map of the island of Eigg was drawn in 1806 and shows the individual buildings, as well as the confused pattern of run-rig cultivation strips in the open fields, frequently interrupted by outcrops of rock and areas of scree and marsh. Shortly afterwards the whole island was reorganized into three single tenancy farms, and crofts. The old run-rig fields with their sinuous boundaries were removed and the twenty or so old buildings in the newly created crofting area were swept away and replaced by seventeen crofts with straight boundary walls running up the hillsides behind the houses. The individual crofts were capable of carrying three to five cows. In addition, there was a piece of common grazing where each crofter was allowed to keep a horse. This area of common grazing has not been ploughed since the reorganization, and the parallel ridges of run-rig, contemporary with the old settlement, can still be seen in the low sunlight of dawn and dusk.

With the collapse of the kelp industry after the Napoleonic Wars, the fortunes of island landowners disappeared and many were forced to sell. The crofters lost their subsidiary source of income, and as the population continued to increase in many communities until the potato famine of the 1840s, poverty increased.[15]

In conclusion, it is clear that with certain notable exceptions, the farming landscape of Scotland was transformed as nowhere else in the United Kingdom, between 1750 and 1840. Here the power of the landlords to create regular fields, to replace hamlet clusters with new farm houses and steadings, to initiate the improvement of land and to build new roads is seen at its strongest, and in many areas this landscape has remained, with few modifications, until the present.

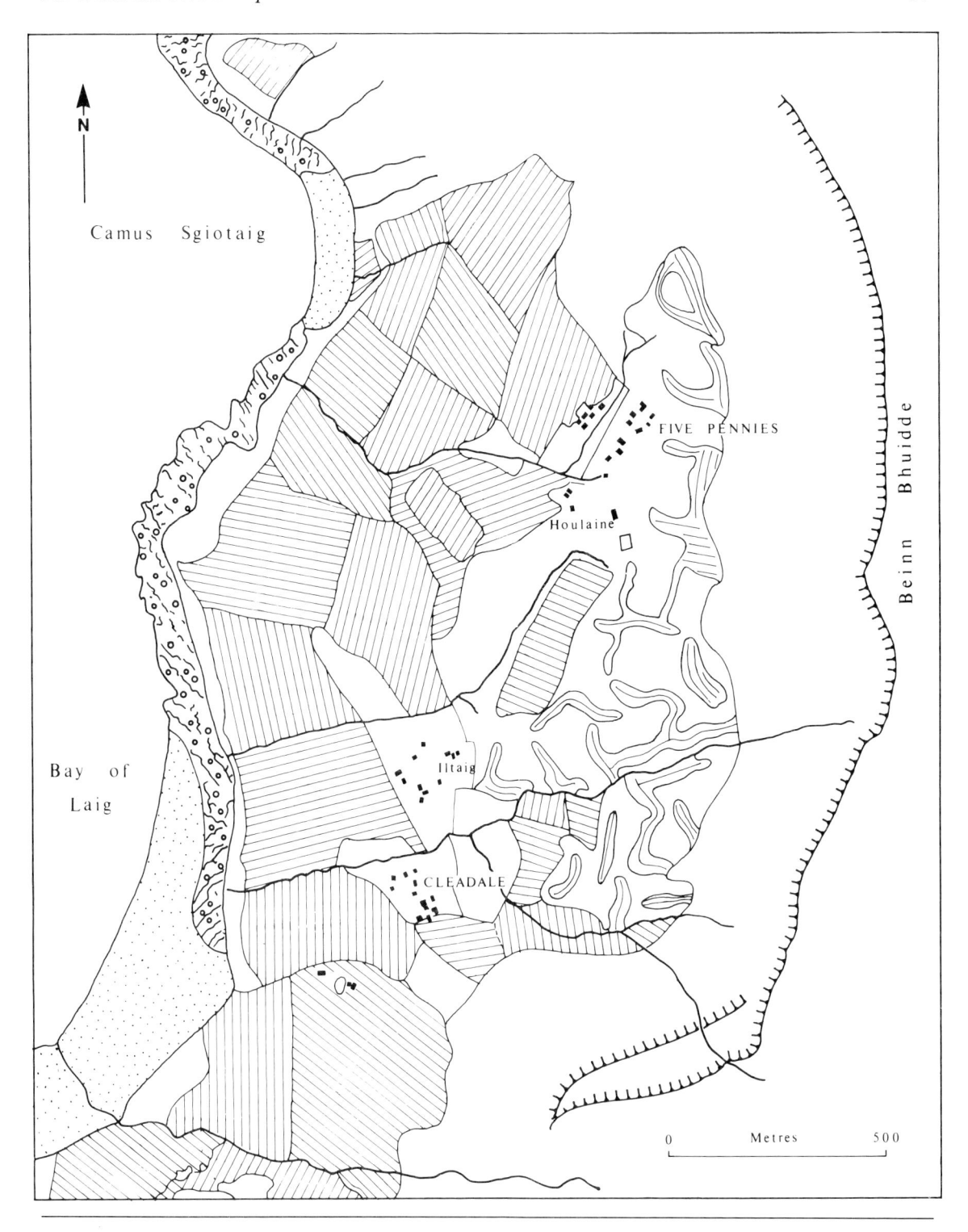

55 *Run-rig fields on the Isle of Eigg (Inner Hebrides) as shown on William Bald's map of 1806*

56 *The northern part of the same area as shown on the 25 inch map showing the walls of the crofts as laid out in the early nineteenth century reorganization, running straight across an earlier pattern of probably very ancient fields (Phillip Judge)*

Summary

The most obvious contrast between England and Wales and Scotland, outside the anglicized areas, is the lack of pre-eighteenth-century nucleated villages. Welsh farming was based on scattered isolated farms, and in south Wales a few longhouses and subsidiary farm buildings can be seen surviving from the seventeenth century. In the North, however, the great estates came to dominate and few older buildings survive. Here, the type of estate farms typical across Britain are to be found.

The best first step towards an understanding of the development of the Welsh farming landscape and its buildings is to visit the Welsh Folk Museum in St Fagans, near Cardiff.

Very little survives in the landscape of mainland Scotland as evidence of farming before 1750, and the best starting place for any research is Roy's map, drawn between 1747 and 1755. It is possible that close study will indicate the survival of a boundary or building incorporated in later 'improvements'. Most of what survives is the result of late eighteenth century and nineteenth century enclosures and the relaying out of farms and villages that accompanied them. The evidence for costly investment in planned fields, farms and villages, dating from the eighteenth and nineteenth centuries is widespread across the Scottish countryside. As profits from sheep dwindled at the end of the nineteenth century the highland estates turned their attention to the sport of deer stalking. By 1912, the deer forests with their castellated shooting lodges and isolated game keepers' cottages had taken over from the farm houses and shepherds' bothies.

9 'High farming' - the heyday of the landed estate

The accession of Queen Victoria to the throne in 1837 coincided with a change in the type of development within the on-going process we call the 'agricultural revolution'.

By 1815 nearly all the open fields and most of the commons had been enclosed and the acreage of grain, encouraged by the high prices of the Napoleonic Wars had increased, spreading across areas of former heath and common which were not cropped again until the First World War. The sands of the East Anglian Breckland and the uplands of north Yorkshire were ploughed up in an attempt to cash in on war-time profits. The war was followed by a downswing in prices and years of depression, with grain prices at their lowest in the twenties and early thirties. The early twenties saw a dramatic increase in farm sales by those who had gone bankrupt or could not pay their bills. Rents on many estates slipped back or remained static through the second decade of the century.

By the late thirties, however, conditions were improving and with the beginning of Victoria's reign came a change of emphasis in farming improvement. It was clear to farmers that grain would never again fetch the wildly inflated prices of the Napoleonic Wars. Particularly after the repeal of the Corn Laws in 1846, a more mixed farming system began to prevail. Grain prices stabilized, and as living conditions in the towns began to improve and railways opened up the urban market, the demand for meat and dairy produce increased. Farmers began to value their animals, not just as manure producers, but as an end product in their own right. Interest in farm improvement was therefore widening from simply the increased output of grain to the whole spectrum of food production for a growing population that was also becoming concentrated in the towns.

Throughout the second and third decades of the century improved breeds of sheep and cattle had spread from the gentry farms to those of their tenants, and specialist breeding was becoming more generally understood. More intensive feeding and housing of stock was a subject in which there was an increasing interest.

The interest of the arable farmers now turned to improved manures, both artificial and natural, and the possibility even of doing away with the break crop between cereals. In this, they increasingly enlisted the help of science. In 1838, the Royal Agricultural Society, originally called the English Agricultural Society, was founded with the motto 'Practice with Science'. In 1834 John Bennett Lawes had begun experimenting on the family farm at Rothamstead and in 1842 took out a patent for the production of superphosphates for fertilizer. The profits from their production enabled him to finance and endow the continuation of experimental work at Rothamstead. In 1845 Cirencester

57 *Digging bush drains at Bedingham farm, Suffolk – an illustration from Rider Haggard* A Farmer's Year *(1899)*

Agricultural College was founded, both to train the sons of tenant farmers and continue research. This was particularly important under Augustus Voelcker, Professor of Chemistry there from 1849.[1] This belief in the importance of scientific knowledge and the breaking down of old prejudices, which so typified the 'golden age' of 'high farming' in the middle of the nineteenth century, was articulated by the Holkham tenant, John Hastings in 1844, when he said, 'Progress will increase as knowledge itself increases'.[2]

This spirit of optimism which permeated through agriculture can be detected in the farming landscape, and it is in the work of the great land owners that it is most obvious.

Although most of the farmed land of Britain was enclosed by 1820, there was still room for improvement, particularly on the heavier soils, and it is the work of drainage with which the middle years of the nineteenth century are associated.

Field drainage

The only methods of field drainage known in the Middle Ages were the open ditches between the ridges in the open fields, and the draining of land by anything else would have been administratively almost impossible in the open fields. The principle of the covered drain was understood by the Romans, but was hardly mentioned in the sixteenth century books of farming advice written by the classically educated Fitzherbert and Tusser.

Walter Blyth, writing in 1649, was the first English author to recommend covered drains. They should be 3 to 4 feet (about 1 metre) deep. The lower level should be filled with faggots of willow, alder or lime covered by a layer of turf, and then 15 inches (40 centimetres) of stone, followed by soil. The trouble with this system was that the drains had only a very short life of between twelve and twenty-five years as the soil above would collapse on to the decaying faggots. Something longer lasting was needed. Hollow brick drains were being made by the early eighteenth century by creating a triangular or rectangular tunnel with bricks. Both brick and bush drains were in general use by the time Arthur Young travelled the country, and their value was well understood.

By the end of the eighteenth century interest in a more permanent type of drain using clay pipes was increasing, and they were being used experimentally by the 1790s. U-shaped brick arches which could be laid on a flat tile were the first to be introduced. James Smith of Deanston, in the Lothians, publicized his own system of drainage which involved using boulders instead of bush, and a subsoil plough which broke up the subsoil but did not bring it to the surface. The cost of such drainage could be as little as £2 14s per acre as compared with £6 14s for tiles. He published his methods in a pamphlet in 1831. He gave evidence to the Select Committee of 1836, saying that the only way to improve output was to drain the land thoroughly.

Interest in drainage continued to increase amongst farmers and landowners alike, but it was clear that there could be no great progress until the invention of a drainpipe-making machine. Pipes were already being used to a limited extent, but they were expensive as they had to be made by hand, by bending a sheet of clay over a wooden cylindrical mandril. The narrow slit left along the length of the tile by this imperfect method was thought necessary to admit the water. In fact the porosity of the pipe was enough. They were first made commercially in Staffordshire and put to general use there on the estates of the gentry by the 1820s. Of the huge and wealthy Stafford estates James Loch wrote in 1820 'an allowance of drainage tiles is made wherever the exertions of the tenants seem to merit such a reward'.

Pipe-making machines were perfected in the 1840s and by 1853 there were forty-five different pipe machines available. They allowed for the cheap production of drainage pipes. Once the general principle of land drainage was accepted, large-scale drainage schemes were likely to be developed.

The real take-off in this later type of improvement began after 1846 when the government, in order to placate landowners over the repeal of the Corn Laws, provided £2 million under the Public Money Drainage Act, and a further £2 million four years later. This money was channelled through five improvement companies which were empowered to lend money for drainage and other improvements. In total £9 million was borrowed from various sources by landlords in the second half of the nineteenth century. This expenditure resulted in the drainage of 4.5 million acres (1.8 million hectares) of land (thirty-five per cent of the total wetland area).[3]

Perhaps it is not surprising that the peak period for investment in drainage was during the time of greatest farming stability, and enthusiasm for high farming between 1840 and 1869. On the Holkham estates it was between 1850 and 1860 that drainage was concentrated, while between 1847 and 1878 the Duke of Northumberland spent nearly £1 million on improvements, much of it on drainage resulting in the recovery of nearly all the wetlands.

In 1855 an enthusiastic reporter wrote of the Durham countryside that tile drains were being laid in every direction, farm offices constructed, and large quantities of fertilizers and manures imported so that 'in a few years the face of the county will be quite altered'. Five per cent was charged to tenants by landlords to cover the cost. The same system prevailed in Cumberland where thirty tileworks making drainage tiles were operating in the Workington area, mid-century. Stone-filled drains were more typical of Westmorland. The Earl of Derby and Lord Sefton were busy draining their

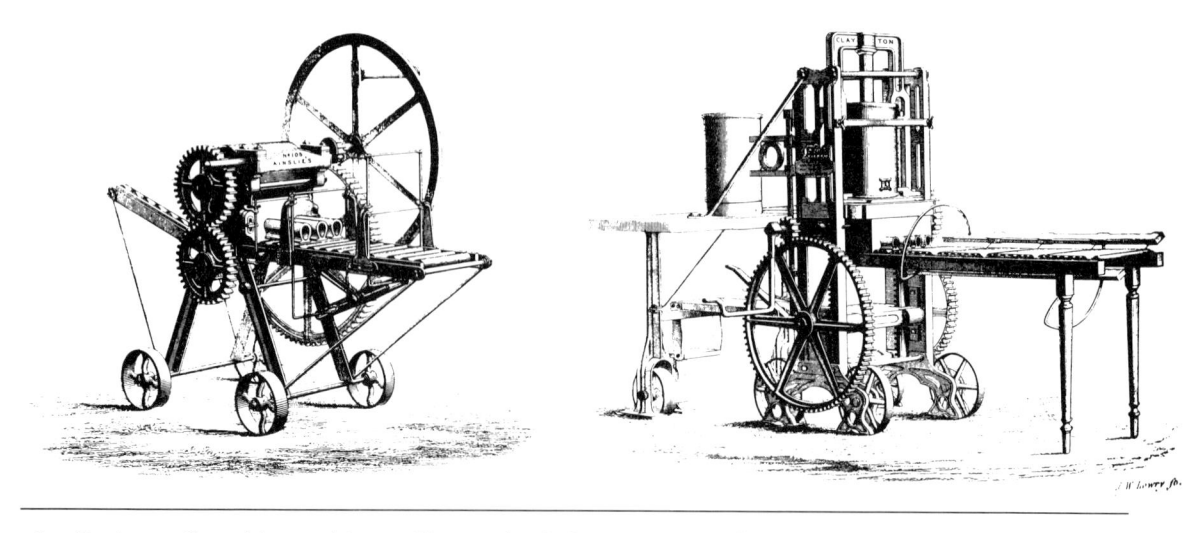

58 *Drainage tile-making machines as illustrated in J.C. Morton's* Cyclopedia of Agriculture *(1855). It was machines such as these that revolutionized the development of field drainage*

Lancashire estates and work was also being carried out in Yorkshire.

Generally there was more draining in the north, midlands and west, excluding Cornwall, than in the south-east and East Anglia, where only fifteen per cent of those soils in need of drainage were affected. It was an expensive undertaking, costing anything between £4 and £8 per acre, often bringing only a small return to the landowner. On many estates five per cent of the cost was charged to the tenant annually on top of his rent.[4] With this low rate of return it is not surprising that drainage was far less likely on small estates of less than 1,000 acres (400 hectares).

There was much enthusiasm for drainage amongst the agricultural writers of the mid-nineteenth century. J.B. Denton, who as well as being a land agent, architect and engineer was the engineer to the General Land Drainage and Improvement Company, wrote that at least a third of cultivated land would benefit from drainage. R.N. Bacon, writing of Norfolk in 1844, was able to find many farmers who fully appreciated the value of tile drains. 'The effect of this system of drainage upon our neighbourhood, says a highly intelligent occupier is that turnips are now generally grown where 30 years ago there was certainly not a swede and but a few whites.' Another correspondent wrote, 'I consider under-draining to be the very best piece of husbandry that can be done on cold wet or springy land, for without it you can grow neither quantity or quality of corn. You may grow straw, but not corn.' On the whole drainage did not result in a change from grass to cereal, but increased areas of roots or green crops. Increase in output was anything between ten and thirty per cent, allowing for an increase in crop value to the farmer of around £1 an acre.[5]

Many of the drainage systems laid down in the mid-nineteenth century remained in operation until the introduction of plastic pipes and massive drain-laying machines in the last twenty years. They were usually laid at a depth varying between 2 and 4 feet (½ and 1 metre) and about 20 to 30 feet (6 to 9 metres) apart. Their effect on the farming landscape was considerable. Wetlands were no longer left fallow and crops were less likely to fail. The production, particularly of more fodder crops, allowed for the

'High farming' – the heyday of the landed estate

intensification of livestock which was an important element in the high-farming system.

Farm building improvements

A significant development at this time, of which expenditure on drainage formed part, was the emergence of the estate as the major contributor to permanent agricultural improvement. We know that 14 million pounds was borrowed for improvements between 1846 and 1882, a quarter of which was spent on building.[6] As well as this there was an unknown quantity of capital generated from agriculture, trade, manufacture and urban land which was available to many a landed family.

The evidence for this, often on a grand scale, is to be seen today on many farms where most of the pre-1940 buildings, except perhaps the barn which may be much older, date from the period 1840 – 70. 'The construction of new farmsteads and the modernization of existing ones was a consuming interest for those landlords who were of an improving mind – they raised farm architecture to the level of a science'.[7] Whilst the most obvious contribution to the farming landscape of the earlier years of the agricultural revolution was enclosure, that in the third quarter of the nineteenth century was farm buildings. Huge home-farm buildings in Victorian 'railway style' architecture grace the entrance to many a park, whilst even the smaller farms gained some buildings and a planned layout during these years. The work of the early years of improvement in the fields needed to be complimented by similar developments in buildings.

Many of the landlords who built these farms did so for much the same reason as they sponsored Sunday schools, 'a feeling of responsibility incumbant upon those of high social standing. Model farms were more a product of Victorian morality than materialism'.[8]

The new buildings of 'high farming' were different to those of earlier years in many ways. The balance between grain and livestock farming was changing and

59 *These model farm buildings at Holkham, Norfolk, are typical of the more ornate, in this case, Italianate, styles reminiscent of railway architecture which were typical of the mid-century*

mechanization was advancing. Farm-building design had to respond accordingly. Grain prices never again reached their war-time levels, but instead, the value of livestock began to increase and so the provision of improved cattle accommodation became more important. Farming was developing from a largely extractive industry to a manufacturing one. The farm was now a factory.

Science was taking its place in agriculture and not least in the analysis of manures. It was becoming clear that manure protected from the weather was a more valuable commodity than one well leached by rain, and this too became a consideration when new buildings were erected. The agricultural writer, James Caird wrote to Sir Robert Peel saying, 'It will be in vain to drain the land and fit it for the culture of

60 *Estates continued to lay out new farms throughout the middle years of the century. This functional set on the Norfolk fen-edge show traditional yards neatly laid out beside the barn. The barn, however, is far smaller than in earlier examples and livestock accommodation more important (Sir Thomas Hare, Stow Estate)*

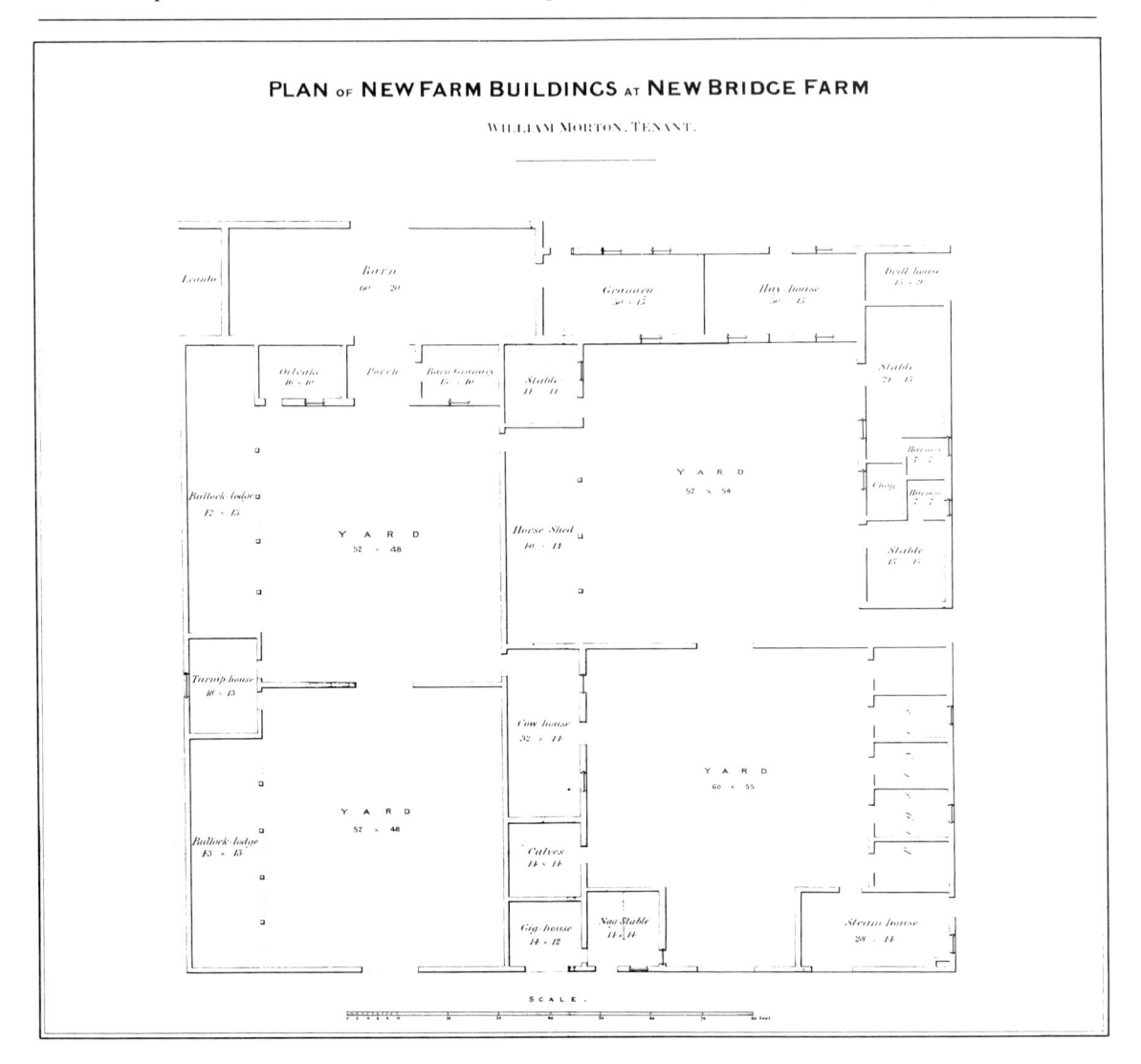

green crops, if no suitable housing is provided for economically converting these into a marketable form and for preserving and accumulating manure'.[9]

Mid-Victorian experts advocated complete rebuilding, but many landlords were not prepared for this expense, variously estimated at somewhere between £4 and £9 per acre, and as such more than the cost of underdraining. Whilst new buildings certainly brought direct benefit to the tenant, the real advantage to the landlord is less clear, although the return was likely to be greater on farms put down to dairying and grazing than under tillage.

Gradually, after 1851, the rural overpopulation which had so long kept standards of living miserably low for most of those in the countryside, began to decline as the industrial towns, and to a certain extent emigration, absorbed the surplus. Between 1851 and 1871 a quarter of a million people left the land. As the labour surplus declined, wages rose and mechanization became more significant. Implement sheds and even blacksmiths' shops became essential elements of the farm plans, and the efficient layout of the farm as a working unit needed to be considered. G.A. Dean was one of that new breed of men, the 'agricultural engineer', who were taking over from architects as the designers of farm buildings. He wrote widely on the subject of farm buildings and recognized the value of housing farm implements which, he wrote, suffered from more wear and tear as a result of being left outside than from actual work. As for efficient layout, he wrote, 'there ought not to be the smallest convenience on a farm, down to a pigsty, that is not so precisely in the right spot that to place it anywhere else would be a loss of labour and manure'.[10]

As the ways of industry began to dominate British life, farmers and their architects began too to think in terms of inputs, processes and outputs and regard their buildings as factories. This is made clear in John Bailey Denton's oft-quoted remark, 'To farm successfully with defective and ill-arranged buildings is no more practical than to manufacture profitably in scattered inconvenient workshops in place of one harmoniously contrived, completely fitted mill'.[11]

The inputs on the farm were more and more likely to have been imported than produced on the farm. 'High farming' involved the end of the 'closed circuit system' of the earlier phases of agricultural improvement. Although there was increased interest in the preservation of the qualities of farmyard manure, artificial fertilizers were being introduced. Superphosphates were available from 1843, nitrates were imported from Chile and guano (seagull dung) from Peru. By the 1860s it was perfectly acceptable for farmers in the grain-growing areas of East Anglia to depart from the four-course rotations laid down in their leases and grow two corn crops in succession, using artificial fertilizers to keep the soil productive.[12]

Animal feeds, also, in the form of cotton and linseed cakes, were available by mid-century and were often fed to stock in large quantities, both to improve the quality of their manure and to fatten them for market. In the 1860s John Hudson spent £2,000–£3,000 a year on oil cake and other feeding stuffs and £800–£1,000 a year on artificial fertilizers for his 1000 acre (400 hectare) Norfolk farm. Imported feed had allowed him to increase the number of his stock from 400 sheep and thirty bullocks, to 2,500 sheep and 150 bullocks. Both the increased quantity of manure and the artificial fertilizers used meant that output of the land was up by a third and the yield of barley had nearly doubled.[13]

The new interest in scientific farm buildings reached its peak in the Royal Agricultural Society of England's competition for farm building design in

61 *Manor Farm, Crimplesham, Norfolk: an example of an 'industrial farm'. The emphasis is on cattle and the steam engine was for grinding animal feed (Field Archaeology Division, Norfolk Museum Service)*

1850. The six prize-winning entries were published in the *Journal* for that year and these show how far farm building design had come since the Napoleonic Wars. First, they had become more 'industrial'. Tramways for moving feed and stacks to the threshing machines were included in many of the plans. The efficient movement of people and products around the plant was everywhere considered.

Secondly, no longer was the barn the most important building. The introduction, first of stationary and then of portable threshing machines that could be worked by a traction engine, had made the double-barn threshing doors and then the barn itself obsolete. A threshing machine might be expensive, but 'the first expense (of a threshing machine) will be overbalanced by the saving in building barns.' Instead of being 'processing shops' where the crop was converted to a saleable commodity – grain – the barn became merely a store, and often its great open space, so essential to its earlier use, was divided up to provide a series of bays for different feeds.

In contrast, the cattle sheds gained in size and significance. In the early 1850s, the cost of stall-feeding cattle through the winter was not recovered by selling fatstock in the spring unless the grain enterprise was charged for the dung. Philip Pusey,

'High farming' – the heyday of the landed estate

President of the Royal Agricultural Society of England in 1840 and Chairman of the Journal Committee, farmed in Berkshire. He tried out many new ideas on his home farm, including the building of water meadows. He went so far as to say that cattle were machines for making manure. By the 1860s, there was a change in emphasis as a result of the coming of the railways, and a gradually improving standard of living. The Norfolk farmer, John Hudson, claimed that whilst before the railway, several days had been spent in driving stock to London, losing an average of 7 pound weight (3 kilograms) in the case of sheep, and 28 (13 kilograms) in the case of cattle, after its opening the journey took less than a day with virtually no loss of weight. By the 1870s, the pendulum had swung to such an extent that grain had to be 'given' manure if cereal production were to be made to pay.

There was no doubt about the increased importance of livestock, but there was much discussion over how to house it. The three methods under discussion were open sheds and yards, covered yards, and loose boxes.

Philip Pusey summarized the situation in 1851, when he wrote, 'Notwithstanding the really excellent plans we (the Royal Agricultural Society) have recently published, I for one certainly should be puzzled; because farm buildings, like certain countries, are really in a state of revolution. Our old ideas about them are unsettled, our new ones undetermined.'[14]

The advantage of box-feeding over open yards was that animals could be treated individually. Even in small yards some cattle would lose out at the feeding trough. 'In a community of a dozen, a few will be tyrants while the rest must be slaves.'[15] Loose boxes began to be built by the 1850s. Dean's farms often included them and some examples of his work survive. Others are found on farms where no architect is known, showing that this type of building was generally accepted in the farming community. They were usually arranged on either side of a feeding passage with a turnip- or cake-house at one end. Many examples are sunken; that is the cattle were led down into them in the autumn and gradually, during the winter, the straw and manure built up so that it was level with the ground outside by the time they walked out in the spring. Providing individual accommodation was obviously expensive and took up space, and so on many farms, yard feeding remained the most important system.[16]

62   *The internal fittings at Crimplesham were not made locally as they would have been fifty years previously, but come from a Birmingham iron foundry*

However, by the middle of the century Augustus Voelcker and others had proved that manure kept twice as much value if cattle were kept in covered yards with gutters. The earliest covered yards to which a date can definitely be assigned are at Eastwood Manor Farm, East Harptree, in Avon, erected in 1858 by William Taylor, who had become a wealthy man by marrying his employer's daughter. He built a grand farm costing £15,000 on his newly acquired 970-acre (400-hectare) estate. The farmstead contained two covered yards with a galleried storage area on the first floor. To the rear of the building was a tramway to move stacks to water-powered threshing machinery. After the water had run the overshot wheel it was led away in pipes to

63 *Interior and exterior views of covered cattle yards at Eastwood Manor Farm, East Harptree, Avon, built in the 1850s, show the optimism and flamboyancy of the period at its most extreme. Again, emphasis is on cattle accommodation rather than cereals; but that livestock were still seen primarily as manure producers for the corn enterprise is shown in the sheafs of wheat which surmount the gable ends (RCHME)*

'High farming' – the heyday of the landed estate

64 *Some of the most elaborate cattle courts were to be found in Scotland. This view of Eastfield Farm, East Lothian, taken from the huge stack yard shows the roofs of the covered cattle court within the outer shell of barn with steam engine, cart lodges, granaries, loose boxes and stables*

65 *Interior of the cattle courts at Eastfield Farm. They were built in 1880 to house eighty cattle. The turnips were stored in the central area and the cattle in pens around the side*

serve an irrigation scheme further down the valley.

Other mid-century examples of covered yards survive across England, mainly on home farms whose often extravagently grand buildings were 'a public proclamation of the status of the landowner and his agent or bailiff in the new world of scientific agriculture'.[17]

These buildings are seen at their grandest in Scotland, where huge complexes of stone- and slate-covered yards, entered through an elegant archway, often with a dovecot above, and usually with a chimney for an engine house behind, were erected on the home farms of the mainly small estates of East Lothian, giving the whole rural landscape an industrial air.

In many upland areas, cattle had long been in-wintered, tied in byres, sometimes on the lower floor of bank barns, and this type of accommodation continued throughout the nineteenth century. Most farmers had to be content with modifications and alterations to existing buildings rather than complete rebuilds. Actual expenditure in a sample area of the midlands showed that the more usual sum was about £3.30 per acre than the amounts of over twice that advocated by improvers.[18]

The monumental buildings of the third quarter of the nineteenth century, if not typical of the average farm, were indeed a very positive affirmation of 'high farming', and the figures produced by some of their owners showed that as a result of very high expenditure on buildings, drainage, equipment, animal feed and fertilizers, profits could be doubled. On J. Mechi's farm, at Tiptree in Essex, the income rose from £5 to £10 an acre.[19] In the long run, however, many of these enterprises went bankrupt and their ingenious tramways, engines and irrigation schemes were left to rust away. It is perhaps significant that John Husdon, for all his expenditure on manures and animal feeds and his support of the philosophy of high farming, never found the need for anything but the simplest buildings on his tenant farm.

These flamboyant buildings may have suited the spirit of the age, but they were not always financially practical, particularly in the period of lower prices at the end of the nineteenth century.

Summary

Whilst it is evidence for enclosure which is the most obvious sign of the changes of the early phase of the landlord's 'agricultural revolution', it is heavy investment in buildings which is typical of this later phase. On most farms some buildings were added, and the type of building is an indication of the changing emphasis of production and increasing intensification on that farm. Often the most ostentatious railway-type architecture is confined to the home farm, and there are many of these surviving across the country, often buried inside an accumulation of modern lean-tos and other additions. This is the first period that power, other than animal power, began to be used on farms and the existence of chimneys, or more often, the stumps of chimneys are an indication of this. It is also the first time that materials other than those locally produced began to be used, and slates found their way into many areas of the country. At the end of the period concrete and corrugated iron made their first appearance.

10 Stagnation and diversification - the last 120 years

The farming landscape of much of Britain in 1870 was as neat and well tended as at any time before or since. True, there was heathland and wetland which might appear today to be 'waste', but even this was cropped for bracken, gorse, sedge and rushes. The fields themselves were hedged, walled or ditched and all these boundaries were carefully maintained. The crops were weeded and the soils boulder-free. There was little fallow in this intensively farmed countryside. Woodland was thinned and trees pollarded to provide the poles needed for fences and hurdles. The buildings were in good order and many had been replaced during the previous thirty years. All these were signs of both prosperity and a relatively cheap and abundant labour supply: both were set to change over the next generation.

'High farming' was a high-input, high-output mixed system. The cost of feeds, fertilizers and expensive buildings were only worthwhile so long as the price of the product, which in many areas was still primarily grain, was high. Until the 1870s the British farmer had few competitors, but after 1870 the steady march of the railways across the North-American continent meant that the fertile virgin soils of the grain-growing prairies were opened up for cereals that could be produced far more cheaply than in Britain. This wheat began to cross the Atlantic and the price dropped by fifty per cent in Britain between the average 1871–5 price and that for 1894–8. The problem was compounded by some very poor harvests; that of 1879 was the worst of the century.

Livestock suffered too. It was not long before refrigerated ships enabled the import of New Zealand lamb and Argentinian beef. Diseases such as liver fluke and foot and mouth were particularly prevalent during these years. However, the imported meat was not of the same quality as home-killed, and livestock farmers did not suffer as seriously as the cereal producers, especially as the drop in grain prices meant that feeds were cheaper.

The collapse of prices forced farmers to cut costs and switch to more profitable crops. The decline in the pool of rural labour had already led to much belated wage rises. The drop in farm income meant that farmers shed workers when they could, and many, especially of the younger men moved to the towns. Inessential tasks were left undone. Hedges were not cut, ditches cleaned or walls mended as often as before. Gradually the countryside took on an uncared for air. Model farms seemed an expensive waste of time and farming practices which were least labour intensive became the most popular. The acreage of permanent pasture rose by twenty-five per cent between 1875 and 1900. With a decline in farming profits went a decline in rents, so few new buildings were erected and the farmer could not afford to maintain those he had.

Alongside the cutting of corners to save labour went the search for new crops that might remain profitable. Sugar beet, which saved farming from depression in the 1930s was introduced briefly in the late 1860s. In 1867 there was a sugar factory at Lavenham, Suffolk, but it had closed by 1872, which was a 'great disappointment' to the local farmers who had received as much as £1,000 a year 'without curtailing the cereal shift'.[1]

There was a general move from cereals to livestock, particularly in areas that were naturally more suited to pasture. The area of arable in Northumberland dropped by forty per cent between 1870 and 1890. In the cereal growing areas, there was a retreat from wheat but still a market for good malting barley and oats to feed the growing population of urban horses. Dairying, particularly around large towns or along the railways leading to them, was the most important growth sector with milk consumption up from 170 million gallons (770 million litres) in 1861 to 600 million gallons (2,727 million litres) in 1900. Essex changed from being primarily a cereal county to a dairying one and the acreage of pasture went up by sixty-seven per cent in the last quarter of the nineteenth century.[2]

66 *Sheep farming has on the whole left little impression on the farming landscape. This circular sheep fold is one of many on the Cheviot hills near Otterburn*

New buildings were needed for these enterprises, which often involved dividing up the great barns that already existed, or building often in cheaper, less substantial materials, such as weather boarding or corrugated iron, which the ever-expanding railway network was making available in even the remotest rural areas. The corrugated-iron Dutch barn was a cheap solution to the problem of storing hay for the dairy herds. Rarely do nineteenth century milking parlours survive. Health regulations since the 1950s have meant that they have had to be altered beyond recognition.

Interest in livestock meant that covered yards continued to be built and open ones to be covered. W.J. Moscrop explained how covered yards had come to be accepted by landlord and tenant alike. 'I had much difficulty in carrying the tenant with me in the first yard I roofed over, but now in letting a farm, almost the first thing asked is 'Will you roof over the yards?'[3] Landlords often found it necessary to erect new buildings, usually for livestock, to attract new tenants or to persuade old ones to stay. Generally, however, the period of expensive building schemes was over and very little was written in the agricultural journals on the subject.

One area where interest in experimentation continued was in the production of silage for feeding, particularly to dairy cattle. Many land owners experimented with enthusiasm using equipment imported from Germany. Experiments conducted at Rothamstead were publicized through an article published in 1886. A Royal Commission was set up in the same year to look into the potential of silage production. Experiments at the Cargan Research Station were published in the *Ulster Agriculturalist* in February, 1888, which looked hopeful. The article concluded, 'The result of this experiment only goes to prove what I have frequently

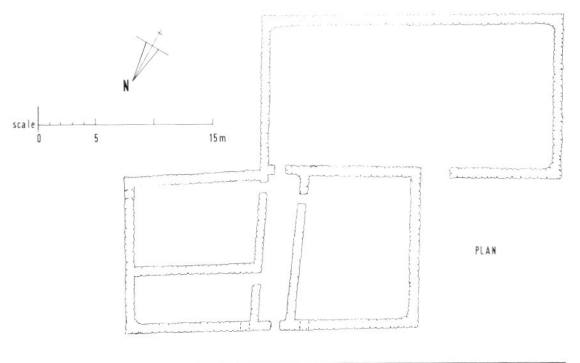

Stagnation and diversification – the last 120 years

67 *More elaborate sheep handling systems known as fanks survive in the Scottish Highlands. This one is at Edderton, Easter Ross (Graham Douglas, Royal Commission on the Ancient and Historic Monuments of Scotland)*

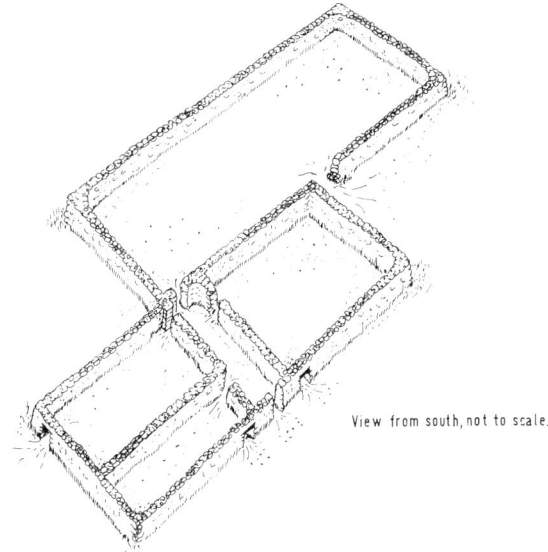

stated before: that inability to make farming pay is due more to ignorance and idleness than to the quality of land farmed.' Cavernous pits under some Northumbrian barns on the Cragside estate of Lord Armstrong testify to his interest in the possibility of making silage, but most schemes were abandoned after a few seasons and there was no real development until the 1930s.[4]

Horticulture was also an important development, not only to provide fresh fruit and vegetables for the towns, but also to satisfy the burgeoning jam, pickle and canning factories. By 1900 a quarter of the acreage of the Fens was used for potatoes.[5]

The period up to 1914 is one of little change in the farming landscape. The capital was not available for land improvement or new buildings and indeed most mid-Victorian buildings remained fossilized until the 1950s. The Gunton estate in Norfolk commissioned a report on the state of its buildings in 1894. This revealed a general state of dereliction resulting from twenty years of neglect. The only way to let farms was to amalgamate them, so the buildings on the out-farm would be abandoned. The land agent who undertook the survey felt that although repairs could have been left to tenants in the past, in the present state of affairs they should be undertaken by the landowner. He was also against farm amalgamation and thought that 'the day of the large farm was over.'

There were some developments, however, especially where money was available from a source outside agriculture. Lord Wantage used his brewing wealth to rebuild farms, and the villages of Ardington and Lockinge in Berkshire.[6] Hampshire landowners, many with London or Southampton dock interests and wealth, continued building throughout the 1880s and 1890s.

There was also experimentation in the use of concrete as a building material. In 1873 the Holkham estate paid £137 to 'Drake's Patent Concrete Company for building apparatus'. An article in the *Journal* of the Royal Agricultural Society of England for 1874 suggested that concrete built up between shuttering could be used for farm buildings; however, it might not be cheaper in areas where brick or stone was easily available. On most estates where it was tried the experiment was short-lived. At Holkham, loose-boxes were built on tenant farms in 1876 and 1883, with several cottages being built between 1873 and 1877.[7] Other examples of concrete work, notably at Buscot Park in Berkshire, bought and developed by Robert Campbell who had made a fortune in Australia, date from the early 1870s. The Buscot buildings are a late example of a 'high farming' type of development. It included a narrow-gauge railway running around the estate to collect sugar beet and take it to the distillery.[8] No concrete buildings were built after 1880 and it was not until the introduction of reinforced concrete in the 1950s that it really became an important material for farm buildings.

The First World War and after

Farming conditions showed no real sign of improvement until the First World War, when, as during the Napoleonic Wars 100 years before, Britain had to feed herself. Land that had not been worked for a generation was ploughed up again and for the first time the government became involved in farming by offering guaranteed prices. Farmers at last felt optimistic and hoped for a better future. When the war ended many landowners were anxious to sell their land before this bubble burst. Forty years of depression had taken their toll on the great estates and debts were mounting. While some were prepared to support the now expensive luxury of land owning, others felt that they should sell before prices fell again. Encouraged to do so by the Liberal reforms, introducing death duties and super tax, many estates came under the auctioneer's hammer. Between 1918 and 1922 a quarter of England changed hands. Many of the farms were bought by sitting tenants and so the whole landowning pattern of the country altered. In 1914 only eleven per cent of England was farmed by owner-occupiers. By 1927, the figure had risen to thirty-seven per cent. This in itself would lead to change in the farming landscape. No longer were the great estate yards with their foundries, carpenters' shops and brick yards responsible for estate buildings and repairs. A certain variety and individualism was likely to creep in as new work was undertaken. The expensive prestige buildings which could only be afforded by men with a great deal of capital behind them were unlikely to be built by the new owner-occupiers. Now it was a case of making do with weather-board and corrugated iron again.

The 1920s and 1930s saw farming again at as low an ebb as it had been before 1914. After the war the government tried to retreat from its role of supporting farming. The price and production of wheat fell to an all-time low in 1931. Land was abandoned. It was often cheaper to leave land uncultivated than try and grow crops. The countryside looked neglected, with weeds and thistles in fields which had recently grown corn. The collapse of the New York stock market in 1929 resulted in a fall by fifty per cent in the price of wheat. This was followed by a new period of government intervention in farming. The various marketing boards were introduced between 1931 and 1933 and the Wheat Act of 1932 at last gave farmers a guaranteed price. However, prices and confidence in farming remained low until 1939. An article in *British Everyman* in 1936 voiced the belief of many at the time when it stated 'Farming is dead in England. Everyone will tell you that!' The farm that Henry Williamson bought on the north-Norfolk coast in 1936 was typical of many.

> Black thistles and the frost-wreckage of nettles lay everywhere on the grass, rotten

sticks and branches lay under the trees, the lane or road leading to the buildings was under water, and the yard we looked into seemed nearly three feet in mud. Flint walls were broken down, every gate was decayed or fallen to pieces. Tiles were off, showing rotten woodwork patched with dryrot. The place seemed entirely forsaken.[9]

A welcome new cash crop of the 1920s was sugar beet, which was sent to Holland for processing. Small quantities had been grown before 1914, but only experimentally. In the 1920s factories were set up in East Anglia and provided a way out of the depression. Sugar beet was a cash crop which could replace turnips in the rotation.

Although machinery was becoming more readily available in the inter-war years, uptake was sluggish and the reduction in the amount of labour employed was slow. Farmers simply did not have the money to spend and had sunk their war-time profits into the purchase of land. However, the seeds of post-war development were just beginning to germinate. Dairying, pigs and poultry were the growth areas and developments in animal housing were taking place. The first battery housing for poultry appeared in 1933 and labour-saving intensive pig housing was also being introduced.[10]

As in the 1880s and 1890s, any new building that went on was likely to be financed by landowners whose money was made outside agriculture as in the case of the white-washed and thatched dairy farm built round a circular yard at Clover Top Farm near Welwyn built by an American millionaire, or the equally picturesque Ovaltine Dairy Farm at Abbots Langley, Hertfordshire.[12] It was dairy farming which was the main growth area and a few dairy farms, sometimes of concrete and often with a silage tower, are the most significant type of new farm buildings dating from this period.

Interest in the use of electricity for agricultural purposes dates from before the First World War, but it was the 1920s that were the main years of experiment. Firstly there was interest in 'electro-culture', a system whereby an electric current was passed through wires suspended above the surface of a field. This induced the crop beneath to grow at an enhanced and more luxuriant rate with consequent beneficial effect upon the yield. So seriously was this taken that in 1918, the Ministry of Agriculture appointed an electro-culture committee which sponsored experiments and produced reports until 1937, when it was finally wound up. Although it was clear that electric currents could increase yields, there were obvious technical problems in setting up the wires and particularly in the depressed conditions of the inter-war years, it was clear that the systen was an economic non-starter. Between 1919 and 1929 Richard Borlace Matthews, an electrical engineer with a farm at Great Felcourt, Surrey, was conducting experiments, not only using electro-culture, but also using electricity to power machinery from ploughs to hay dryers. As a result he only needed to keep two horses on his 600 acre (240 hectare) farm. The evidence for much of this activity is still intact on his farm.[11]

That agriculture had in general stood still for fifty years is shown in that unique record of the farming landscape, the RAF aerial photographs of the 1940s. They show the complete post-enclosure hedged landscape as it had been created by 1830 and before it was devasted by modern developments. A quarter of the hedges (120,000 miles (192,000 kilometres)) were to disappear between 1946 and 1974. Usually it was the more ancient small fields rather than the large rectangular nineteenth century enclosures that were vulnerable to destruction. The loss has been greatest where the hedges no longer have a function, so in the livestock areas of the west, such as Cornwall, more ancient hedges survive than in arable East Anglia. It was not until a high proportion had disappeared that the significance of the loss was appreciated.

The food-producing campaign of the Second World War really began the changes in the agricultural landscape that have continued at an increasing pace ever since. 1939–45 saw the introduction of mechanization and fertilizers. Horses were slow to disappear; there were still 300,000 in 1950, but they were a rarity by 1960. With them went their grazing and the fields of oats, allowing for the more intensive monoculture which has removed much of the visual interest from the farming scene.

During the Second World War the Scott Committee was given the task of formulating future land policies. It laid down that 'self-sufficiency in food should be a primary objective of post-war agricultural policy', but failed to appreciate the great changes about to overtake farming, which are sometimes seen as another agricultural revolution, altering the landscape as fundamentally as any that had gone before. The Agriculture Act of 1947 ensured that in the future agricultural change would be influenced by government policy.[13] This was increased after 1973 with Britain's membership of the EEC.

Since 1947 the race to produce more food has resulted in between a third and a half of our ancient woodlands being lost, and sixty per cent of fens drained, whilst heathlands have been reduced to mere fragments. Small farms, particularly in the hills, have been declining in number at a rate of two per cent a year. The size of farm that is needed to support a full-time farmer has nearly doubled every decade since 1945. The result therefore, is fewer, larger farms, containing regular and often huge fields in an almost hedgeless landscape.

The increasing wealth which government support has brought has allowed farmers, for the first time in 100 years, to replace their buildings. Pre-stressed concrete was developed in the 1950s and was immediately taken up by the farm-building construction industry. New machinery, ever-increasing in size, needed larger sheds. Intensive animal husbandry meant larger buildings and the erection of huge chicken and turkey houses.

These generalizations can be backed up from the regional studies carried out by the Countryside Commission in 1974. They show the situation at the time of Britain's entry to the Common Market. Many of the changes described below were to accelerate thereafter.

The most obvious alterations to the landscape were the loss of hedgerows and the erection of large multi-purpose, mass-produced buildings, but the degree of change varied from area to area.

Although there were few hedges to remove in the Cambridgeshire fens, dykes were filled in as larger pumps meant fewer main drains were necessary and more drainage could be underground. The average size of a field in 1974 – 31 acres (13 hectares) – was more than double that of 1945 – 14 acres (6 hectares).[14]

In Huntingdonshire, on the heavy midland clays, an area that had traditionally concentrated on livestock became intensive cereal land. This meant that economies of scale were more necessary and resulted in a massive loss of hedges. Average field size in the sample area rose from 19 to 45 acres (7 to 18 hectares) between 1920 and 1970, while the number of holdings had halved, with the average size rising from 122 to 317 acres (49 to 128 hectares) over the same period. Most of the changes would have been post 1940.[15]

The light Dorset chalklands, however, had been enclosed in the late eighteenth and early nineteenth century and so the fields had always been large. Although there was a swing away from sheep to cereals after the war, fewer hedges were removed.[16]

Where agriculture remained primarily for

Stagnation and diversification — the last 120 years

68 *These drawings show typical changes in the flat Cambridgeshire countryside over the last fifty years. Hedges have gone as fields have been enlarged (*New agricultural landscapes*, Countryside Commission, CCP76)*

livestock, as in the dairy farming areas of Somerset and the heavy clays of Warwickshire, farm size has remained more stable and fewer hedges have been lost. Sometimes, where farms were amalgamated and the system completely rationalized, old hedges were replaced by more flexible wire fencing systems.

By the late 1980s, government and EEC policies, which had encouraged increased output, particularly of cereals, so that Europe could be self-sufficient, had resulted in massive surpluses. In the early 1990s there have been efforts to persuade farmers to reduce output. This, coupled with an increase in private transport and leisure time, and the public awareness of the pleasure of 'country pursuits' has resulted in government and EEC efforts to encourage farmers to produce less and diversify. Some farmers now find golf courses more profitable than agriculture, and the farming landscape is set to change again. Areas of woodland are likely to increase as more tree planting is undertaken and some hedgerows replaced. Set-aside policies have resulted in uncultivated fields for the first time for forty years. Unlike the 1930s, however, British agriculture is now highly capitalized and mechanized. In spite of moves towards organic farming most of the area that remains in cultivation is likely to be as intensively farmed as ever.

As a result of the momentous rethink of agricultural policies that is still being undertaken, the farming landscape will undergo further fundamental changes over the next ten to twenty years, but whatever these are, they will simply be the latest phase in a history of continuing development, going back more than 5,000 years.

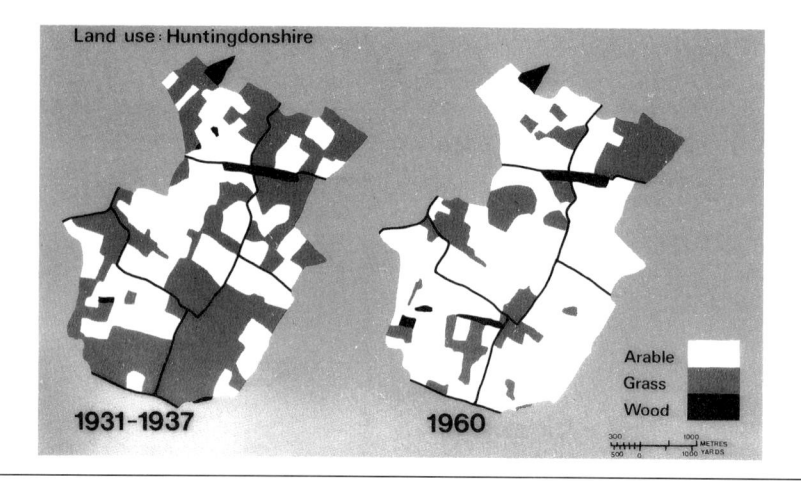

69 *This diagram shows the changes in land-use in parts of Huntingdonshire between 1937 and 1960. The most striking change is the increase in arable (*New agricultural landscapes, Countryside Commission, CCP76*)*

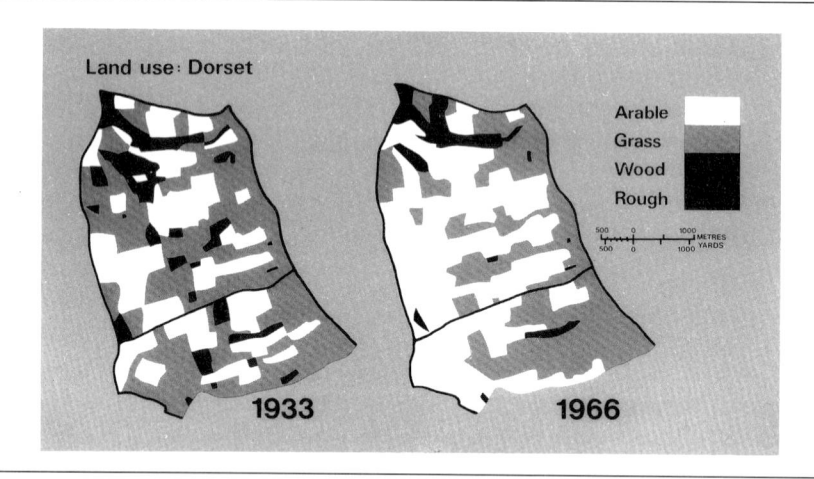

70 *Changes in Dorset have been considerable, but more grassland survives in an area that is naturally less suited for arable farming (*New agricultural landscapes, Countryside Commission, CCP76*)*

Summary

Periods of agricultural depression are times when there is little deliberate change in the countryside. Fields are unlikely to be altered and only the minimum of buildings are put up. This was the case during much of the period considered in this chapter. Some landlords put up a few buildings, usually cattle sheds, in order to keep tenants, while a few wealthy newcomers to land could still afford more ostentatious buildings. The main growth area was dairying and the most important technological breakthrough was perfecting the making of silage. The most typical farm building dating from this period was therefore the dairy farm, often an iron-framed concrete group of buildings, one of which was a silage tower.

The changes of the last forty years have been far more dramatic involving the destruction of much that had gone before, both buildings and field boundaries, and the erection of prefabricated wide-span structures for grain storage, machinery and livestock, which are themselves now becoming obsolete.

THE RECLAIMED LANDSCAPE

11 The reclaimers

So far this book has described the gradual development of the farming landscape under a variety of human, economic and social influences. At some periods, change was faster than at others, stimulated by population increase and high prices for agricultural produce. This is true of the creation of sheep walks at the end of the Middle Ages, and the enclosure movement of the Napoleonic Wars, but even then, change was generally small scale and gradual, building on what had gone before.

There are, however, some areas of the United Kingdom where change was far more dramatic, resulting from the influence of often no more than a single family, or institution and its agent. There are many estates where the hand of one landowning family can be seen at work. This is as true of the great monastic estates of the Middle Ages as the neatly enclosed fields with farm houses and cottages conforming to a standard estate design of the 100 years after 1750. It is here that we see the most tangible evidence for the influence of the landed estate on the progress of the 'Agricultural Revolution'.

By the late eighteenth century the only uncultivated areas left in Britain were either inhospitable moors and uplands, or ill-drained fens. It was these that offered greatest scope for improvement and, it was hoped, quick profits. Faith in science was such that improvers believed that there was no limit to the extension of agriculture if only enough money and effort was put in. As a result, the first seventy years of the nineteenth century was a period when more expensive and unprofitable schemes were undertaken than at any other time, and it was only the very wealthy who could undertake the most expensive.

The reclamation of much of the Yorkshire wolds was the work of Christopher Sykes of Sledmere. The rediscovery of the art of building dew ponds made it possible to keep stock on the dry uplands. This was followed by enclosure, and the creation of new farms and the building of farmsteads, with as many as fourteen being under construction at one time in 1788.[1]

Northumberland, much of which was unenclosed wasteland, had been divided into large farms between 300 and 1,200 acres (120 and 485 hectares) by 1820 as a result of the work of landowners such as John Grey of Dilston. Threshing machines were adopted early and buildings improved. Sir John Graham consolidated his three hundred and forty Cumberland farms of about 90 acres (36 hectares) each into sixty-five larger units in the 1820s. Redundant buildings were pulled down, old irregular hedges removed, and drainage channels cut.[2]

The Royal Forest of Exmoor was sold to John Knight in 1820. All over 800 feet (240 metres) above sea level, this empty area of moor had supported nothing much more than deer and rabbits, and the remains of artificial rabbit warrens, known as pillow mounds, survive in front of the aptly named Warren Farm. John hoped to farm Exmoor as one huge arable farm. Deep ploughing and liming, he hoped, would make the soil suitable, but his plans took no account of climate and when this proved impossible, his son, Frederic, divided the estate into mixed livestock farms, eleven of which survive today. Surrounded by rectangular fields with hedges mounted on stone walls, and backed by a shelter belt of beech, these characteristic farmsteads can be seen across Exmoor. Roads were built and a new parish of Simonsbath was created in 1856, with a church to serve the new community.[3]

The reclamation of fen, although expensive, was more likely to be successful than the efforts on the high moors where not even the most optimistic Victorian could change the climate. Parts of Chat Moss, Lancashire, were drained in the early nineteenth century, and the descriptively named area of Sunk Island on the Humber estuary was drained by the Crowns' Office of Woods in the 1850s. S.S. Teulon was employed to design cottages, a school and a school house, which were built, whilst his church was not. More work by this somewhat eccentric but popular Victorian architect can be seen at Thorney in Cambridgeshire, described in the following chapter.

The two chapters that follow deal with some of the most expensive improvements in Britain, undertaken by two of the most wealthy families, at opposite ends of the country. The Dukes of Bedford had long been famous for their efforts in the fenlands of Cambridgeshire, but it was not until the 1850s that their estates at Thorney were successfully drained and took on the appearance with which motorists using the A 47 through the fens are familiar today. The Dukes of Sutherland in the extreme north of the Britain did not begin to transform their estates until they had available the enormous wealth of the Bridgewater canal. Their 'improvements', involving the clearance of small tenants from much of the county, have long been infamous, but the new farming landscape they created has received less publicity.

These two areas have been chosen because the great wealth of their owners enabled them to set in motion a wide range of expensive schemes almost uninhibited by financial constraints. In the long run neither undertaking paid dividends, but both have left a lasting mark on the countryside and are a reminder of the lengths to which man has gone to bring new lands into production.

There are, however, many areas of reclaimed landscapes that could have been singled out for discussion, and the activities of estate landlords are worthy of study both on the ground and through their archives. Both field and documentary work need to be carried out side by side. Whilst maps and documents help in an understanding of the landscape, the landscape too helps in the interpretation of the documents. If either the documents or the landscape have been destroyed then a vital dimension is lacking for the study of the development of the area.

Only rarely is the documentary record complete, but a selection of papers from most estates survives, either still in private hands or in the county record office, and these sources need checking first. The most useful are likely to be maps, estate surveys and descriptions, often prepared by a new agent when he took over the mangement of the estate, or sometimes commissioned from a firm of land agents, perhaps when the estate changed hands. Leases, account books and letter books are also valuable. Letters can be particularly useful when the owner is an absentee landlord and all the details of estate mangement are reported to him in writing. Finally, if the estate was sold, as many were in the years immediately after the First World War, the sales details describe each farm and its buildings in glowing terms for prospective buyers.

Other papers that may be useful include enclosure maps, the tithe maps of the late 1830s and 1840s, the county reports of the Board of Agriculture made between 1790 and 1812, and the county reports commissioned by the Royal Agricultural Society of England and mostly published in their *Journal* in the 1840s and 1850s. (For further details on documentary sources see *FARMING Sources for Local Historians* by Peter Edwards (Batsford, 1989).)

It is only by visiting a chosen area of study that flesh can be put on the bare bones of the story. It may be that old field boundaries have been swept away and farms amalgamated, but where buildings do survive their existence establishes that they were in fact built, and that schemes which appear on paper were carried out. Their survival is an indication of how suited they were to local farming conditions and details of function and design can be studied.

Whilst it is hoped that the two studies which follow are of interest in their own right, a second aim is to encourage further study of the many other estates across Britain, where a wide variety of less famous landlords were in their own way making a vital contribution to the reclaiming of agricultural land across the British Isles.

12 The Duke of Bedford in the fens

Sources

Sources available for this study included firstly the printed information to be found in the various books on the drainage of the fens[1], and one written by a previous Duke of Bedford describing the running of the estate in the nineteenth century.[2] Some of the farms were illustrated in John Bailey Denton's book, *The Farm Homesteads of England*, published in 1863. The first edition Ordnance Survey 6 inch map, published in 1887, showed the area and the farms shortly after reclamation and farm building was complete.

Documentary sources are mainly to be found in the Cambridgeshire Record Office amongst the records of the North Level Drainage Authority, and include plans of the estate, steward's letter books, registers of tenancy agreements, rentals, valuations and a commonplace book in which the steward recorded the day-to-day running of the estate.

All the farms of the Thorney Level were then visited. Many have disappeared in the last fifty years; some have been demolished with amalgamations, whilst others have fallen down because of the weakness of their foundations on the shrinking peat. One of the most valuable parts of visiting the farms was talking to the farmers, particularly those who had been there for several generations and who were able to give details of traditional farming and the management of the estate.

The Duke of Bedford owned 18,500 acres (7,500 hectares) of land in the Cambridgeshire fens around Thorney. Thorney Abbey was granted to the first Earl of Bedford in 1550 and the area is unique in the fens in that since then it has been owned by one great landowner for nearly 400 years.

The most useful land was about 300 acres (120 hectares) of 'high ground' around the abbey itself. This island of gravel rises to about 25 feet (7.5 metres) above see level, and the line of this old enclosure can be seen in the dykes today, bounded by the old Thorney River and the Ten Foot Drain. The boundaries within the old enclosed area are far less regular than those across the rest of the parish. Outside this area the entire estate was below sea level and described in 1574 as 16,000 acres (6,500 hectares) of fen 'formerly dry, but now waterlogged.'

The problem of draining this area north of the River Nene, known as the North Level, was made worse by the silting up of the river below Wisbech. What was needed was that it should be straightened and embanked from Wisbech to the sea. Without this, any efforts at improving the Nene flood plain inland from Wisbech were difficult in the extreme. The urgency of the situation was brought home to the merchants of Wisbech when ships were finding it difficult to reach the port, and work was begun to dig a new channel in 1719. However, the project

The Duke of Bedford in the fens 125

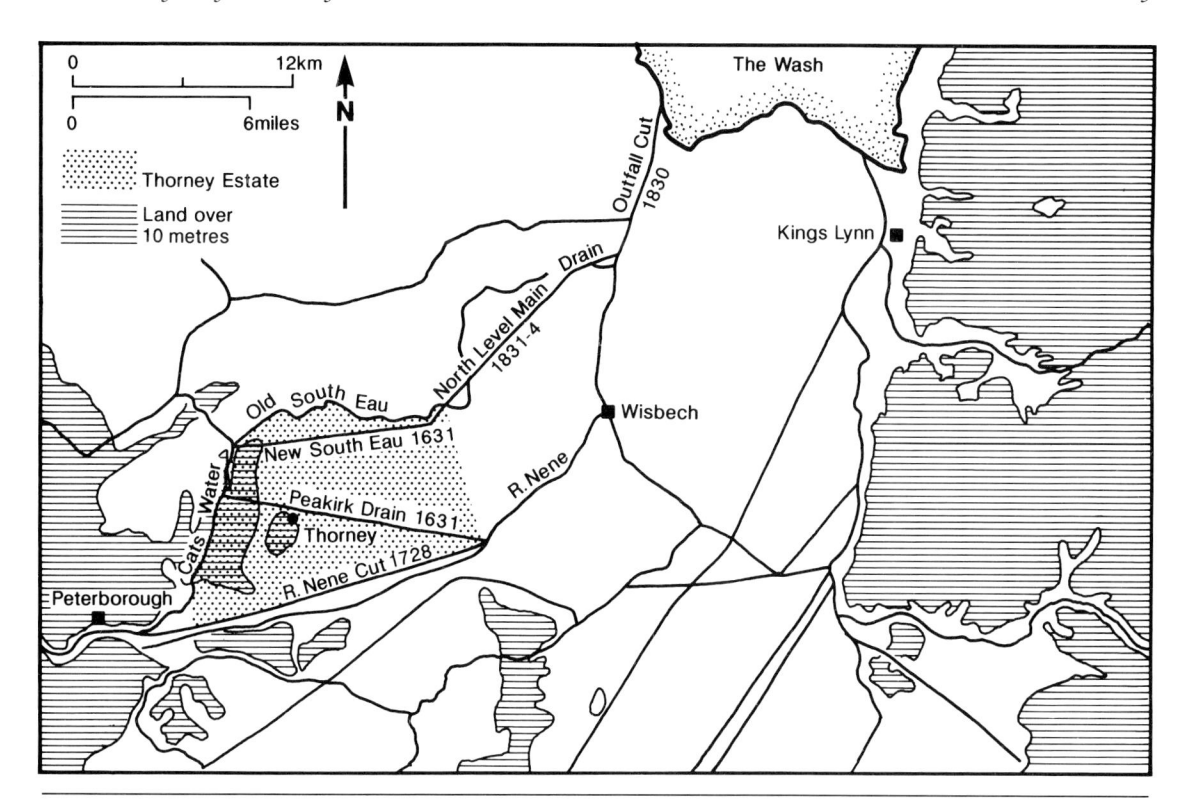

71 *Location map, Thorney Level (Phillip Judge)*

proved to be more expensive than anticipated and was abandoned.

Some efforts were made in the parish of Thorney. The channel of the River Nene, which forms its southern boundary, was recut and straightened in 1728, and new sluices put in. To the north a new channel for the South Eau was cut in 1631 and later new banks thrown up to stop it flooding. By 1750, sixteen wind-powered pumping engines were working in Thorney in an attempt to keep the parish free of water.

In spite of the problems, the parish supported a large population in the eighteenth century, with about 300 tenants listed in a rental of 1700. Many of the tenants' names were either French or Flemish, a reminder of the importance of the French and Walloon settlers. Many came to work on the seventeenth century drainage schemes and stayed. They had their own church services and baptism registers where, between 1654 and 1663, an average of twenty-nine births a year were recorded. During the next ten years, numbers rose with an average of nearly thirty-eight. By the late eighteenth century, numbers were dropping as the incomers were assimilated into the local population. However, the two droves to the north of the village are still called French Drove and English Drove, a reminder of the previous divisions in the population.

The total population of the parish remained fairly constant throughout the eighteenth century with the figure of 250 families being quoted by Charles Vancouver in his report on *The Agriculture of Cambridgeshire*, published in 1794. Many of the farm names which appear on later maps

72 *Thorney Level, with its regular pattern of farms and ditches. The irregular layout of 'highland' around Thorney village, cultivated since the Middle Ages, stands out from its surroundings (Phillip Judge)*

are also listed, showing that this population was already widely scattered across the fens and not concentrated in the village.

The 1760s and 1770s were years of floods, and pressure from the inland farmers on Wisbech mounted so that finally, in 1773, work on the Nene outfall began again. Kinderley's Cut below Wisbech, lowered the water level of the North Level by 5 feet (1.5 metres) and improved the navigation up to Wisbech. Problems were, however, far from solved and Knarr's Cross Farm, beside the River Nene, was described in 1782 as 'a very bad farm, liable to be overflown.'

Charles Vancouver described the agriculture of Thorney in some detail in his report of 1794. Firstly, there was the clay and gravel highland area covering about 500 acres (200 hectares), let at twenty-five shillings an acre and used for permanent pasture and grazing. Then, the first quality peat fen covered about 3,000 acres (1,200 hectares) and could be rented at about eighteen shillings an acre, again best used for grazing. A further 6,000 acres (2,400 hectares) of second quality land was let for fifteen shillings an acre and was used for the cultivation of wheat, oats and coleseed, and then put under temporary pasture. The poorest fen made up about 8,000 acres (3,200 hectares) of the parish and could be let for no more than eleven shillings an acre. Here, oats and coleseed could be grown and then the land put down to temporary pasture.

Surprisingly perhaps, on an estate with a reputation for improvement, such as the Duke of Bedford's, long leases were not granted and farms were held 'at will'. But in spite of this, 'a spirit of improvement pervades the mind of every tenant beyond what is to be met with in any other part of the country'. Courses of husbandry were, however, stipulated. On the two poorer types of fen, the old temporary grass was firstly pared and burnt, 'but with great care and under proper limitations', then coleseed was sown and fed off to sheep. This was followed by a crop of oats, followed by one of oats or wheat. The land was then returned to pasture for not less than six years and grazed with sheep.

73 *Thorney Level in 1749 (Cambridgeshire Record Office, R77/38)*

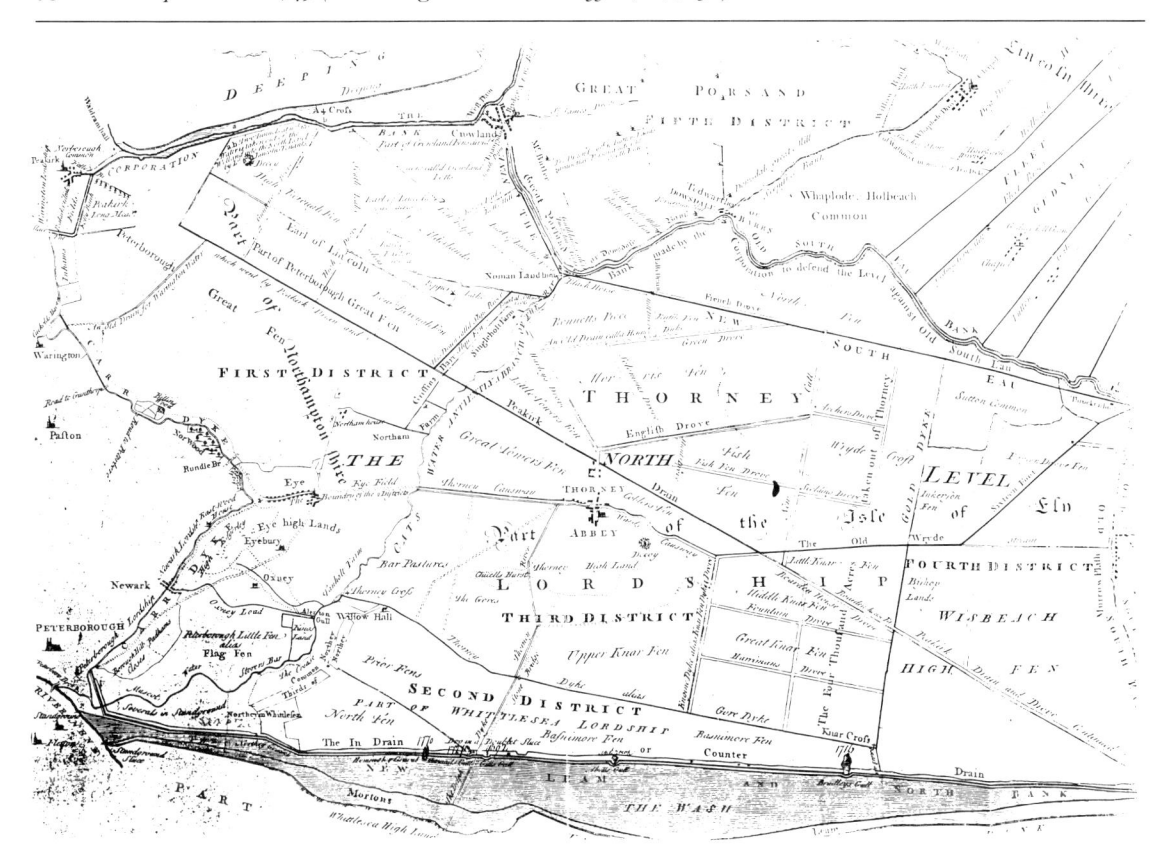

By 1800, the drainage problem had increased again. Great sand banks were accumulating in the Wash below Kinderley's Cut and the Bedford Level Corporation was pressing for action. Sir John Rennie was commissioned to report on the Wisbech outfall and his report was published in 1814. Thomas Telford was also consulted and the result was the building of a new outfall, following acts of parliament in 1826 and 1827. Once the Nene Outfall had been lowered by 10 feet (3 metres), the water began to flow in the drains in Thorney. The North Level Main Drain was built between 1831 and 1834 and 'land that could have been bought 20 years ago for £5 per acre was now worth £60–£70'.

The Duke of Bedford, through his agent, Tycho Wing, took a great interest in drainage across the North Level, of which Thorney formed only a part. Wing wrote a pamphlet on the problems, published in 1820, and supporting Sir John Rennie's scheme. As the windmills could not lift water more than 4 feet (1 metre), they could only work when the water level in the Nene was low, at low tide, and so were idle for ten to twelve hours a day. Added to this was the fact that they could not work in a gale. The clearing of the outfall and the installing of a steam engine was thus important. The benefit of this would be that spring-sown corn could be sown earlier, the wheat crops would not be lost, the grass seeds would not be washed away, and the pasture would not be spoilt with twitch which thrived in wet conditions. 'The best husbandry may be adopted with confidence of success.'

After the instalation of steam power, the cleaning of the outfall and the completion of the North Level Main Drain, Wing's confidence was indeed boosted to such an extent that he wrote, in 1834, that now the new drainage was in full operation and 'acts in the most preferred manner now that the steam engine is working well', he would pull down all the windmills. He was offering seven or eight redundant mills to his neighbour whom he heard was about to undertake some new drainage. He wrote to the Duke of Bedford at Woburn, 'I am preparing to take down and sell as fast as I can dispose of them, the windmills formerly used for drainage on this estate . . . it is impossible that they can ever be required again for the purposes of drainage.'

The opening up of the port of Wisbech again meant all kinds of enterprises could be contemplated. One such scheme, seeking the support of Tycho Wing, was a paddle-steamer service to London, to carry cattle and sheep, thus saving the loss of weight suffered by the animals when they had to walk. The area was certainly prime sheep grazing country when Cobbett rode through in the 1830s and commented on the fat sheep as well as the good yields of wheat and oats produced without any further manuring.

However, the work of drainage was by no means finished. The North Level Commission continued to be active, although always short of funds. The silting up of the outfall was a continual worry and the banks needed maintaining. The channel of the Nene, above Wisbech, was being widened in the 1840s whilst the embankment at Sutton Bridge was being strengthened. Sir John Rennie, junior, was a consultant to the engineering works and visited the area in 1849. He had not been in the fens for over a year, but 'was gratified to see farm buildings, roads and the whole under cultivation.' Wing replied, saying, 'I am very glad you have seen our new marshes and banks as well as within and without and an instructive lesson it is as to what perseverance can do: but the laurels are all yours.'

Drainage on the Thorney estate continued to be a problem, particularly in the very wet years of the 1880s, but once the North Level Main Drain had been successfully completed, the estate was encouraged to carry on with its own works, and expenditure rose through the 1840s and 1850s reaching a peak in 1861 when £19,326 was spent in one year on 'new works and permanent improvements.' By 1875, the end years of agricultural prosperity, £188,991 had been spent since 1839, creating a landscape which makes Thorney distinctive within the fens.

This work included not only the main drains, but also underdraining of the fields using clay pipes. Wing was of the opinion that the best way to do this was to follow the 'one-only' rule, which meant the landlord did all the work and charged it to the tenant. Farmers, however, preferred to use their own men, as drainage provided winter work, but this could be solved by the estate taking the men on. 'If I want to drain 100 acres in the parish, I go to each farmer on whose land the work is being done and enquire how many draining men he can spare me.' A gang of twelve men under a single superintendant could then move round and carry out the work. Tenants' enthusiasm for the work was increasing and in a single fortnight, in 1843, 100,000 drainage pipes were used.

In 1849, the flamboyant church and country house architect, S.S. Teulon, was employed to design cottages in the village and his Victorian Tudor terraces come as such a surprise to drivers along the A47 as they approach Peterborough from the east. 'It is not enough to produce good order on a nobleman's estate by driving out all the bad and dirty,' wrote Tycho Wing. He must also set a good example in cottage design. Thorney was set to become a 'model community in an otherwise disorganised and unhappy farming area', and as a result we have probably the most extensive group of well-designed Victorian cottages in England.

However, Teulon's work did not go without criticism from the agent. Of one group of twelve cottages the agent wrote to Teulon complaining that they were too

74 *The elaborate estate workshops at the tankyard, Thorney*

spacious. 'Small children normally sleep in the same room as their parents, whilst the bigger ones want fetching out into service at an early age for if retained late under their parents' roof, idleness and profligacy are too frequent results . . . and the separation of the family into three comfortable bedrooms may be rather more condusive to this than otherwise.'

It was not only cottages that were rebuilt, so were farm houses and buildings and there is no doubt that Teulon was also employed to design several farm houses. These yellow brick and slate substantial Jacobean-style houses, several of which survive in the parish, may well have been his work.

It is not clear how many of the eighty farms in existence by 1880 were on pre-nineteenth century sites. Early farmstead sites avoided the peat and were built on the silts and gravels of old river courses. Early eighteenth-century houses using stone, presumably robbed from the Abbey, survive at Harriman's Farm, Park House, Willow Hall, Hill Farm and Prior's Farm. Of these, only Park House and Hill Farm are on the high ground. The farms on the 'highland' of Thorney, as well as some along Thorney causeway (the A47) and the drove parallel to Catswater, which includes

Willow Hall, mentioned in the seventeenth century, are all old sites. Others, spaced regularly along the straight droves, were probably not created until after the channel of the River Nene was recut in 1728, or perhaps as late as the 1850s. Because of the unstable foundations on peatland, few old buildings survive and many have been replaced again and again. Houses are mentioned on all the fens in the late seventeenth century, usually with barns and sheds beside them, but their sites have probably moved several times since then.

The most impressive farm buildings to survive were those designed by the agent, Robert Mein, to serve farms varying in size from 150 to 450 acres (60 to 180 hectares) and some of which are illustrated in J.B. Denton's book, *The Farm Homesteads of England*, published in 1863. They were built of the yellow brick produced in the estate brickyards in Thorney (to a size different to

75 *Mr Goodman's farm, Thorney, designed by the estate manager R. Mein and published in J.B. Denton's book* The farm Homsteads of England *(1863)*

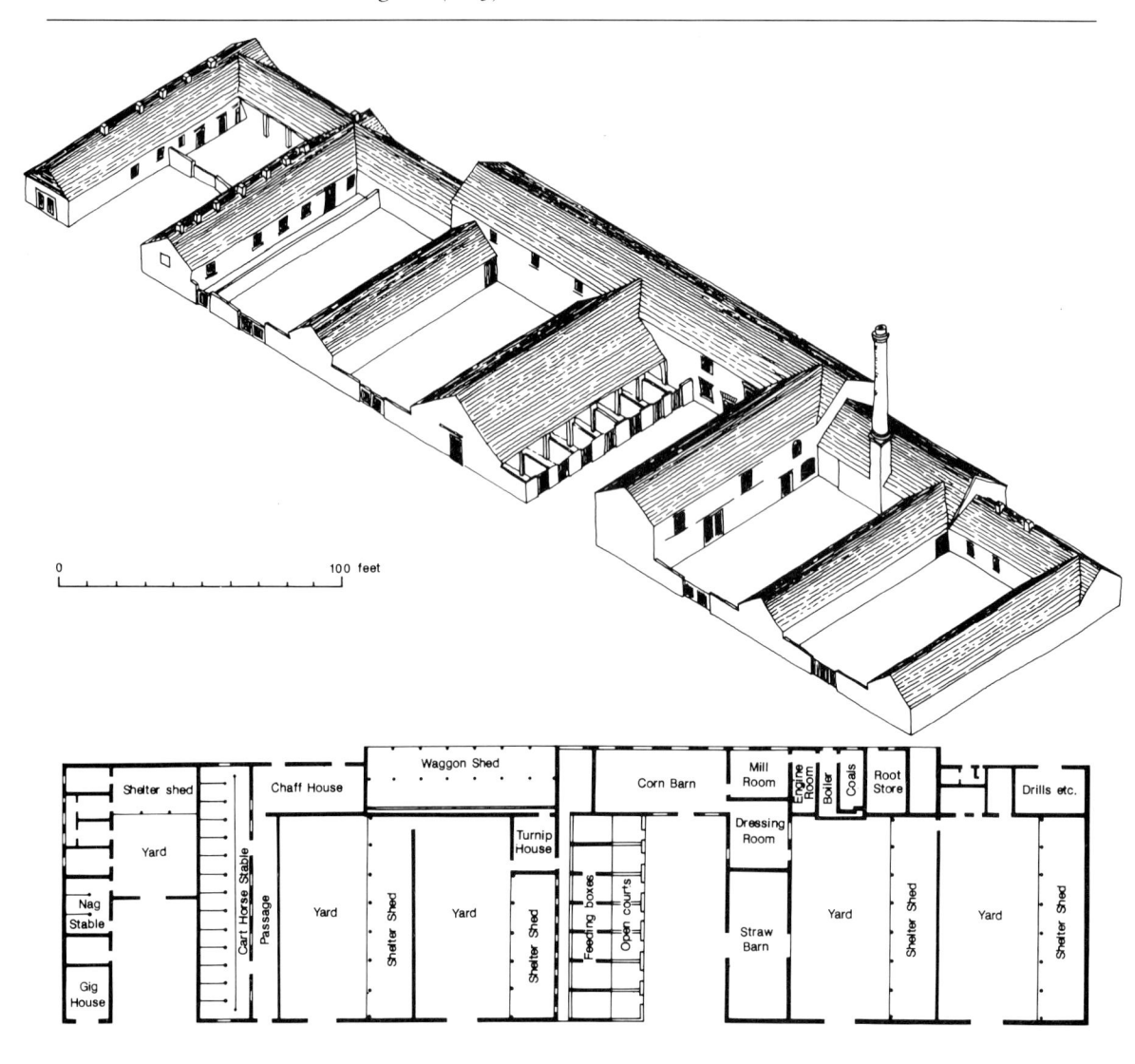

that standard elsewhere), roofed in slate, and boarded inside. Two different designs are illustrated. On both the barn is a relatively insignificant building, little higher than the cattle yards, of which there are four, each for ten beasts. There is also a row of loose boxes with a feeding passage for individual fattening. Cart-horse stables are at one end of the range, providing stabling for between ten and fourteen horses. The larger farmstead has a nag stable and trap house and a steam engine to work a threshing machine in the barn. Examples of both these types of design can still be seen standing in the fens although no steam engines or chimneys survive. Typical of these farms are the cast-iron window frames and sliding doors. It is not clear how many farms were rebuilt at this time, but by the 1880 first edition 6 inch Ordnance Survey map, there were at least twenty farmsteads on E-shaped plans, whilst those shown on an estate map of 1857 are still of an irregular layout. These impressive buildings were often too substantial for their sinking foundations and had to be replaced. By the 1880s weather boarding was being used instead of brick, as a lighter building material, less likely to subside on the unstable peat. The buildings were described in an estate survey of 1881 as being good and adequate, but more cattle yards and sheds were needed on outlying fields.

Although it is clear from the surviving buildings, estate surveys and map evidence as well as the figures for improvement quoted by the Duke of Bedford (£263,675 on 'new works and permanent improvements' from 1836–95) that much money and effort went into turning Thorney into a model fenland estate, little effort was made to enforce improved farming methods through leases. Tycho Wing was very scathing about the tenantry. 'It is the most intelligent men who are the most difficult to deal with, for they are the people who attend least to the practical carrying out of labour . . . they are away from home more, leaving no one in charge and although they can check on above ground work, they can't check below ground.'

Until the early nineteenth century, sheep were the most important product of the fens, although Vancouver noted that crops were grown on the land less suited to grazing. However, further underdraining meant that by the 1860s arable was gaining in significance and the new ranges of buildings built show that cattle, too, were important. Husbandry leases date back to the seventeenth century and were included in tenancy agreements. It is surprising that on what was well-known as an 'improving estate', the course of cropping imposed in the nineteenth century was a five course one in which a fifth of the land was to be left fallow in any one year. A further fifth was to be under beans or clover and the rest was to be used for a white corn crop, but not for more than three years in succession. The idea of leaving land fallow and growing three grain crops in succession seems a very primitive one for the period.

76 *Knarr's Cross Farm, Thorney was completed in 1865. This view along the yard shows the shutter and window design typical of the estate*

77 *A small farmhouse on the Thorney estate*

By the 1880s, the majority of the land was cropped, with nearly 12,000 acres (4,900 hectares) arable and 7,000 acres (2,800 hectares) pasture. However, the wet seasons of the late 1870s had shown up the problems of farming much of the wetter land. More underdraining was obviously needed. On Bukehorn Farm, for instance, the land was very low and 'has suffered much from wet and yields an uncertain return.' At Knarr Corner 'the lower parts of the farm have frequently drowned in recent years and has suffered in consequence. The tenant is an industrious man and but for adverse seasons would have thriven.' The problem of drainage in wet years was not really overcome until the more powerful engines and improvements of this century.

It was not only the weather, but also the fall in grain prices that was undermining farming prosperity, and many farmers were described as lacking capital, so it was not surprising that a general rent reduction of seven per cent was recommended in 1881.

The drainage of the fens and the creation of the modern fenland landscape was a long process, but one in which the Bedford estate, active on its own estates at Thorney, played an important part. Now the land is divided amongst several owners, farms are amalgamated and old houses and buildings derelict or demolished. But the signs of the work of the Dukes of Bedford are very much part of the fenland environment today.

13 The transformation of the county of Sutherland

Sources

Whilst improvements had been taking place in the lowlands of Scotland from the mid-eighteenth century, it was not until the nineteenth that they reached the remote parts, and nowhere was this more dramatic or publicized than on the Sutherland estates of the extreme north. Here an entire region of nearly a million acres (405,000 hectares) was transformed at the direction of one family and under the control of a single estate office controlling local factors.

The estate received a great deal of publicity from the time it began its policy of 'improvement' and there have been many books on the subject, the most useful of which for this study was James Loch's justification of estate policy[1], published in 1820. The most comprehensive modern study is that by Eric Richards[2], which was written using the documents in Stafford Record Office. The papers from the Sutherland estate for the period 1802–16 have been published by the Scottish Record Society[3] and so the descriptions of the early improvements could be studied from a printed source.

The bulk of the Sutherland papers are now in the National Library of Scotland (Dep. 313 and Acc. 10225) and were consulted there. As well as maps of the area before and after improvement, and plans of some of the farms, there were surveys, letter books, field books showing what was being grown in every field of the individual farms in the 1830s and 1840s, lists of improvements and allowances made to tenants for their improvements. There was, in fact, a huge quantity of documents and only a very few of those available were consulted.

Due to its size it was impossible to cover all the estate, but all the farms for which there was a record were visited, and changes in estate policy and the type of farming which was being promoted could be followed through investigation of the buildings.

The beginning of improvements

Complete reorganization is an expensive business and was only possible first because of the marriage in 1785, of Elizabeth, Countess of Sutherland, to George Granville Leveson Gower, Lord Stafford, who owned extensive estates in Shropshire and Staffordshire, forming one of the most productive land holdings in England benefitting from its position on the edge of a rapidly expanding industrial area. Secondly, in 1803, another source of income became available to the family with the death of the Duke of Bridgewater. By his will the interest from his highly profitable canals, trading enterprises and some landed estates

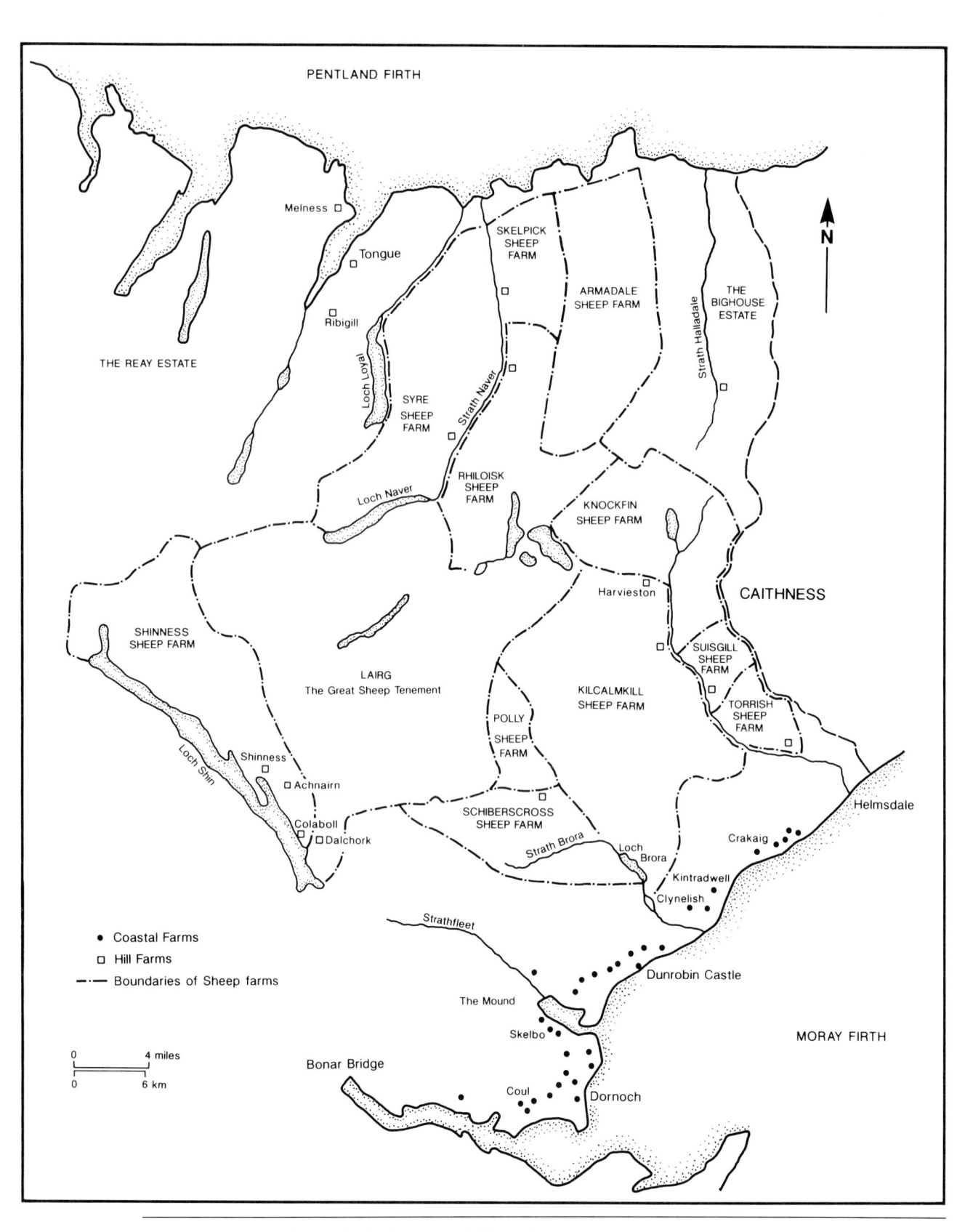

78 *Location map, the Sutherland Estate (Phillip Judge)*

was to go to Earl Gower. This title was inherited by the Countess' eldest son in 1805. Between 1803 and 1816 an average of £64,500 was paid out every year by the Bridgewater trustees to Earl Gower. Although much of this was used in Staffordshire, part of it also went north, and the period of improvement can be dated from July 1805, when the Countess and her son went north to inaugurate the new era of investment.

The 'spirit of improvement' which inspired so many landowners in the early nineteenth century could now range unrestrained by financial worries, and the first twenty years of the nineteenth century saw greater changes in this area than at any other period, as nearly a quarter of a million pounds was poured into Sutherland. Many costly, often ill-thought out, and hasty schemes were embarked upon over the following years, and it became a feature of all these projects that costs would exceed expectations.

The 'pre-improvement' landscape

Much of the estate to which the Countess brought her husband and son was wild, bleak and inhospitable moorland, but cutting through this upland were the wide straths, or valleys of Kildonan, Fleet, Halladale and Naver, where, in the shelter of the hills, crops could be grown and animals pastured. Potatoes, bere and black oats were grown and in the words of the agent there had been 'no attempt at improvement.' Along the coastal strip, too, particularly between Bonar Bridge and Helmsdale, there was good arable land and much more that could be brought into cultivation if the winding rivers from the straths could be canalized and land drained.

By 1800 all this cultivable land was densely populated. Roy's map of Scotland, surveyed between 1747 and 1755, shows the settlements along the straths, and in some areas there was an almost continuous band of run-rig. This would have been farmed in a traditional infield-outfield system, but as the population rose and there was more pressure on land there was less chance for the outfield to remain fallow. The result was that land became less fertile and crops less reliable. The number of years in which famine was a real possibility increased. Mass starvation was only avoided by generous distributions of meal by the landlords. This duty would not have seemed so arduous to the landlord in the days when he had wished to keep up population to ensure recruits for his private army, but after the last Jacobite rebellion of 1745, when a more peaceful regime settled on the highlands, land was not valuable for the number of people it could support so much as for the rent it could command. It soon became obvious that the most profitable use of land was to clear it of inhabitants and let what had been townships in multiple occupancy to single tenants either, in the case of the uplands, as huge sheep farms, or on the coast, as well laid out mixed holdings. It was easy for the Countess and her factors to persuade themselves that this was for the general good of the people. Descriptions of the conditions in the straths in which the adjectives 'squalid' and 'wretched' are frequently used, abound. One observer exclaimed that living conditions were 'scarcely fit for a wild indian.' There is no doubt that continuous cropping was reducing the productivity of the area. The Malthusian belief repeated to Earl Gower, the Countess' son, by one of the factors in 1810, that 'where nature can not provide food for the numbers, she sometimes suits the numbers to the food', would certainly have seemed very near the truth.

Archaeological and map evidence can add something to these rather subjective views of life before the clearances. One clearance settlement at Rosal in Strathnaver was excavated in advance of forestry and is now a site managed by Historic Scotland.[2]

The settlement was surrounded by about 80 acres (32 hectares) of arable land which was separated from the surrounding moor by a dry-stone wall. Not all the land within the wall was cultivated; at least a third was rough pasture, and in spite of all the problems, soil evidence suggests that the area had retained its fertility well. Within the enclosure, the remains of about a dozen houses can be seen. These were of a longhouse type with a byre at one end and living accommodation at the other. Other buildings in each group included a barn and often a yard. The stone walls were not more than 3 foot (1 metre) high. Above the stone the wall was peat. These 'mud' and stone hovels inhabited by the farmer, his family and animals are fully described in the reports of the period, and these accounts would appear from the archaeological evidence to be correct. The roof was supported on a cruck or 'couples' which were recessed in slots in the stone walls. As they were usually cut from the woods of the proprietor, and were therefore his property, they were left when the house was vacated, whilst the cross timbers were bog wood and therefore belonged to the tenant and so were taken. The roof would have been turf and the hearth was in the centre of the floor. The main fuel was peat and the amount of time spent gathering it was a major complaint of the improvers. If farmers spent all the summer cutting peat how could they devote the necessary attention to their farms? Far better that they should burn coal and use their horses on the land.[3]

Maps of the settlements shortly before they were cleared were drawn for the estate between 1813 and 1820. They show the pattern at Rosal to be typical of other places: a scatter of dwellings within a walled area as well as other buildings which were either barns or for general storage. Many of the houses had small enclosures beside them and most settlements included a kiln for drying corn and sometimes a mill for grinding. These simple and primitive settlements were on sites often occupied from prehistoric times. At Rosal there is both a bronze age burial cairn and iron-age souterrain within the enclosure, and it is this continuity that was swept away in the clearances.[4]

It is clear from some surviving maps of the area in Dornoch parish, south of Loch Fleet, that some improvements had taken place here by 1790. A map of Skelbo shows newly laid out fields and a courtyard layout described as 'barn' beside two long buildings described as 'farm offices'. Maps of nearby Coul show planned steadings beside scattered buildings. In the case of Easter Coul, the new group is described as 'farm steading' and the scattered one as 'old steading'.

In the case of Skelbo, these changes would have taken place while the estate of Skelbo was alienated from the Sutherland ownership, and it did not return to them until 1808. Whether Coul was part of that estate is not clear.

Early nineteenth century change

The main period of change was, however, the first two decades of the nineteenth century. Roads and bridges were built, a mine sunk and lime burning initiated. Coastal fishing stations were set up, negotiations for industrial undertakings, such as flax mills, begun and the major task of buildings a dam across Strath Fleet to keep out the sea and allow for the reclamation of land behind, as well as providing a roadway, was completed.

Between 1805 and 1816 the main trunk highway along the east coast (now the A9) was built and the road to the north coast at Tongue was partially constructed. At the same time a start was made on a system of local roads and bridges. Many of these early bridges are still in use, with their distinctive parapets with rounded ends. Although road building was a joint venture with the Commission for Highland Roads and Bridges, much Sutherland money and initiative went into their construction.

Nowhere was this more true than in the building of the 'mound' or dam at Strathfleet. In 1811 William Young, one of the factors, came up with the idea of carrying the road across the Fleet on a causeway, partly to ease transport, but partly to make possible land reclamation behind. 200 acres (80 hectares) could be reclaimed from the sea and 108 acres (44 hectares) of meadow be protected from the tide. Telford, the Commission's engineer, was sceptical and there were problems in finding a contractor. In the end Young and his fellow factor, Patrick Sellar, undertook to do the work. The first problem was moving the necessary quantity of material. Sellar was confident that a hundred barrowmen and fifty-four men using twenty-seven waggons on a horse railway could move the necessary weight in one season. Work began in March 1814 and part of the causeway was in place by the winter. The real problem was closing the final gap and several eminent engineers, including John Rennie, were consulted. In the early summer of 1816, 700 men with 250 barrows, fifty carts and forty waggons were assembed to complete this section which had to be done as quickly as possible to prevent destruction by the tide. Finally, on the 26 June, after many problems, the Marquis and Marchioness crossed the mound on their way to Dunrobin. The whole project had cost nearly £10,000.

The coastal farms

Other work was more directly concerned with the improvement of agriculture. Much of the flat coastal area needed draining if it were to be farmed on improved principles. Drainage tunnels, sometimes as deep 6 feet (2 metres) were dug and then lined and roofed with huge slabs of slate and flag. Many have survived up to the present, but are now being broken by the weight of modern machinery, allowing the immensity of the original drainage schemes to be appreciated. Elsewhere, the winding rivers and streams were canalized and often utilized to provide power for the threshing machines installed in the new farms.

The new farms which were surveyed and laid out were of two types. Those for the single large tenants who were to replace the communal holders of previous years, and those for the small tenants who were to be moved to the coastal areas and settled on smallholdings consisting of not more than 2–4 acre (1–2 hectare) strips.

The signs of all this activity are obvious in the county's landscape today. Although the A9 has been greatly improved and a new bridge constructed, it still crosses Loch Fleet beside the Mound along which the old road ran. Upstream is the reclaimed land protected from the sea; downstream are the sand and mud flats of the Fleet estuary, and the salmon leap between the two. Along the coastal strip, between the Mound and Helmsdale, is evidence of the major farming improvements which took place in the first twenty years of the nineteenth century. In 1800 most of the area that was cultivatable was in run-rig, with the land split amongst sub-tenants under a chief tenant or tacksman, the area around Kintradwell being typical. The map of 1772 shows cultivated strips, 'heath' and 'improvable land', as well as a scatter of buildings, often within their own enclosures. The land described as improvable was probably in need of draining and the map shows several little streams apparently disappearing into pools and lochans. The old lease on this land was not due to run out until 1818, and until then it was 'entirely occupied according to

Kintradwell 1722

Cragganmore

Letoch

Balaunroich

Improvable

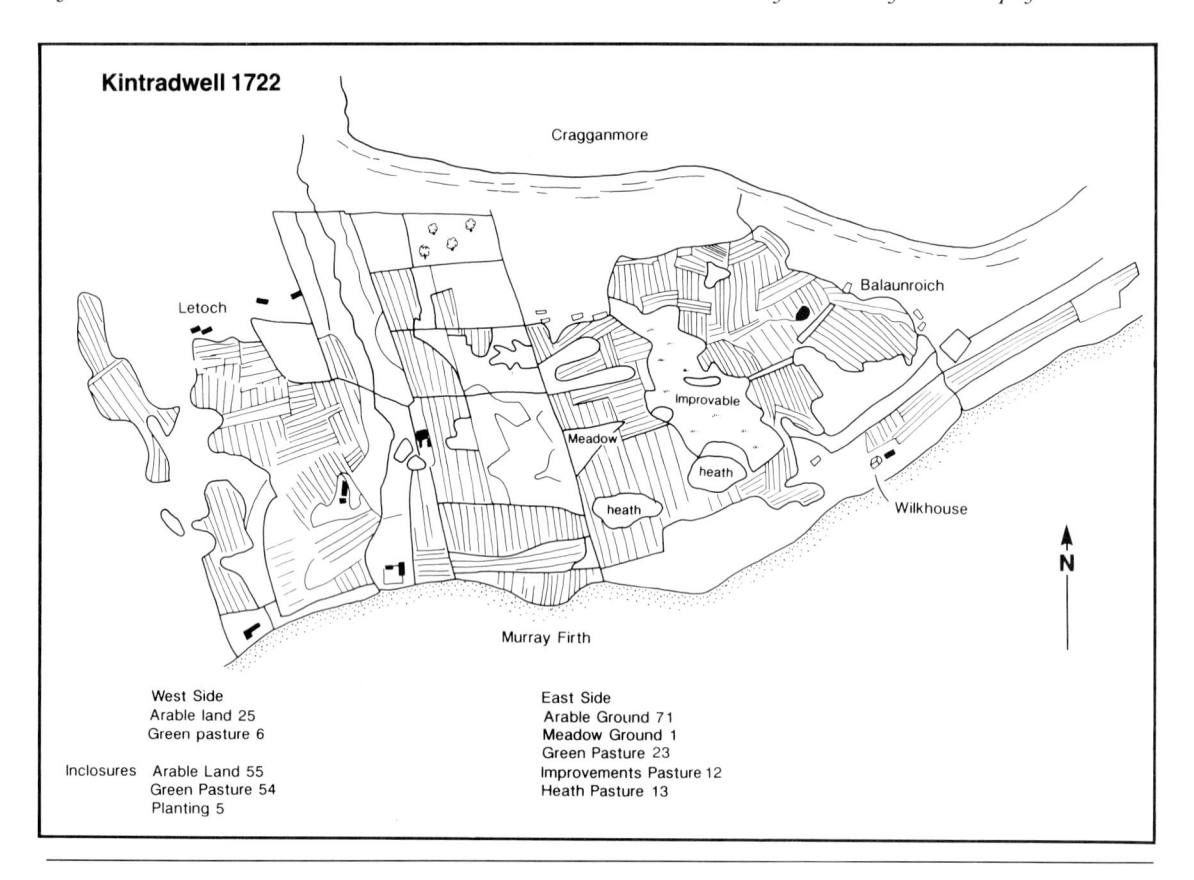

Meadow

heath

heath

Wilkhouse

Murray Firth

N

West Side
Arable land 25
Green pasture 6

Inclosures Arable Land 55
Green Pasture 54
Planting 5

East Side
Arable Ground 71
Meadow Ground 1
Green Pasture 23
Improvements Pasture 12
Heath Pasture 13

79 *The township of Kintradwell in 1772 (redrawn by Phillip Judge from a map in the National Library of Scotland)*

the old method. In 1819 the old sub-tenants were moved to new holdings in Brora, the farm divided into large rectangular fields and new farm buildings erected. The outline of those buildings still remains although there was a certain amount of building work in 1862 and the yard is now covered. Like most of the farms of this date, the farm consisted of a courtyard, with an open shelter shed and a granary above along the back, byres and stables down one side, a barn with a water-powered threshing machine along the other side and a wall along the front, in this case with a couple of piggeries. Beside the steading was a bothy. Typically, the 1862 alterations involved the building of a central wing of loose boxes and a new bothy.

The first farm to be reorganized had been that of Culmaily. The problems encountered there were typical of those on many others. The sub-tenants were difficult to move and the tacksman, too, refused to allow access until the last minute. Finally the tenants were allowed the ground for building the new steading, but the tacksman would not release the mill and waterfall on Culmaily burn for the threshing mill to be built, 'the root of all our future doings'. The buildings were completed, the last tenants removed and the tackman compensated and pacified by Whitsun, 1810, when the new lease on the farm commenced.

Gradually, all along the coast, the old communal tacks (or leases) ended, and by 1820 the small tenants moved out. The land

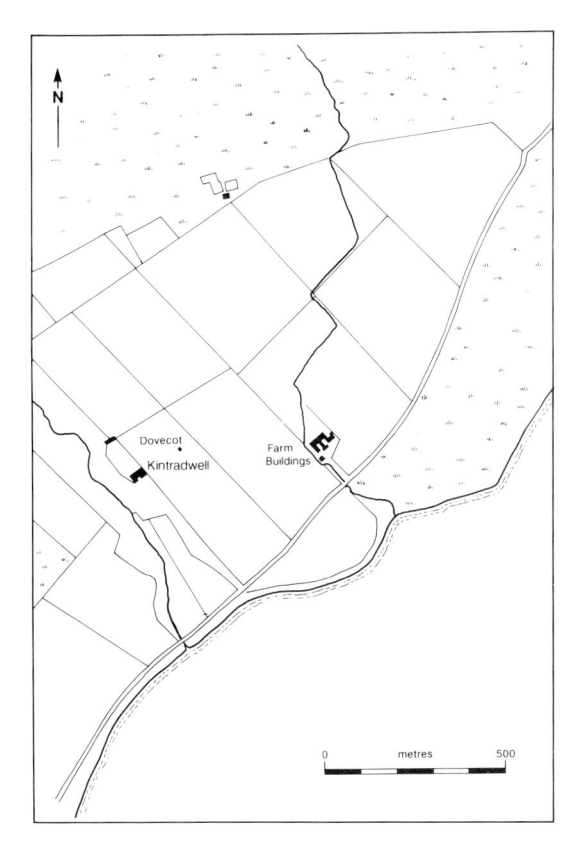

The transformation of the county of Sutherland

was laid out in large rectangular fields and miles of dry-stone walls were built, drains dug and buildings erected. Fifteen or so planned steadings were built along this 20 mile (32 kilometre) stretch of coast. Some included new houses, but others incorporated fine older houses, presumably the residence of the former chief tenant. Where new houses were provided there were well-appointed dwellings for a comfortably-off tenantry, with elegant staircases and carefully proportioned rooms. Only rarely is there evidence that old farm buildings were incorporated in the new. In none of the barns, for instance, is there any

80 *Kintradwell in 1880 (Phillip Judge)*

81 *Kintradwell farm buildings as shown in James Loch's report (Phillip Judge)*

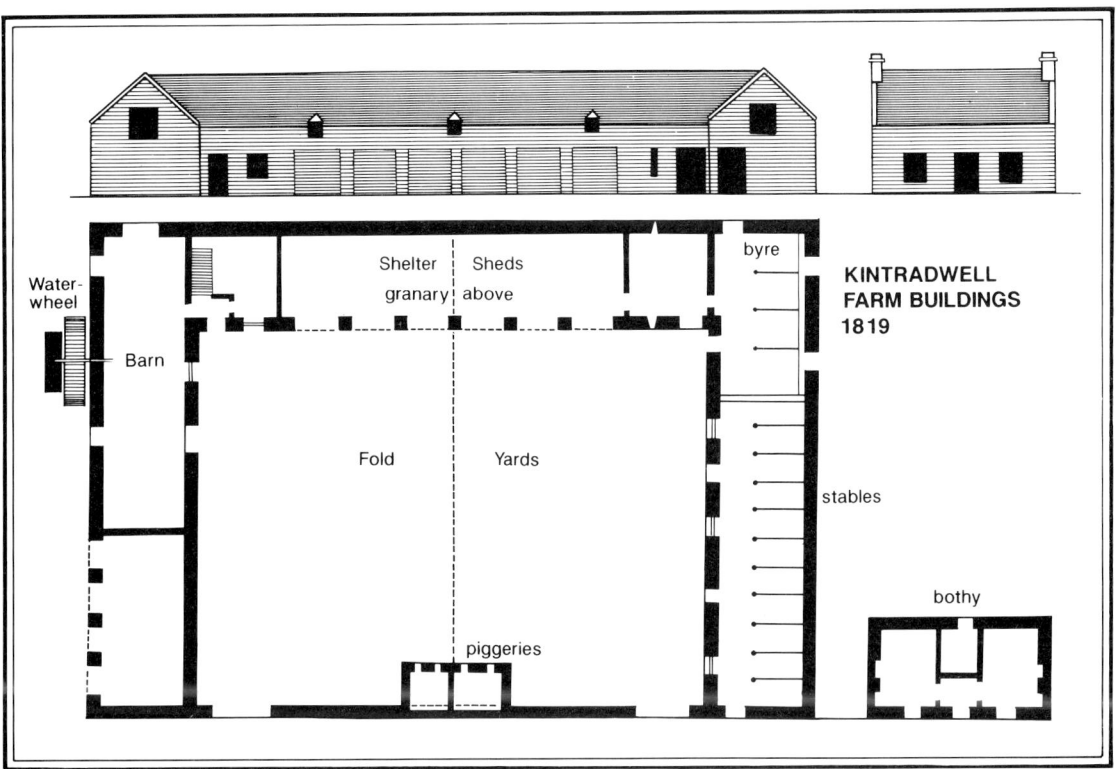

KINTRADWELL FARM BUILDINGS 1819

sign that they ever had double-threshing doors for hand flailing before they were adapted to water power. Some of the houses and buildings have the Stafford/Sutherland coats of arms on the date stone.

Many have been altered since they were originally built. At Kintradwell livestock accommodation was added, loose boxes replaced shelter sheds and sometimes yards were covered. In a few exceptional cases, as at Clynelish and Skelbo, completely new steadings were built in the second half of the nineteenth century.

Over the entire estate thirty-six lowland mixed farms mostly, between 150 and 200 acres (60 and 80 hectares), were created and laid out. In 1829 plans of these were drawn and for the next forty years the crop rotations practiced on them recorded. This gives us a unique record of farming methods in the mid-nineteenth century. Although the principles of improved farming were clearly being observed, in that two grain crops were never grown in succession, the system was biased towards hay and pasture. Whilst one might expect between a third and a quarter of the land to be under grass, here it was anywhere between a third and a half. Potatoes were an important crop, occasionally replacing turnips in the rotation, and beans were also sometimes grown. Barley and oats were the most important cereals with rye and bere less usual. Wheat was only rarely cultivated. This pattern is confirmed in a letter from the factor, William Young, to Lord Stafford in 1810, 'In these districts turnips, flax, barley, clover and oats with occasionally a little wheat will of all others pay best.'

Terms of leases show that it was usual to leave land under grass for at least two years. Five and seven year rotations, sometimes including a year of fallow, were not uncommon. Even under careful management, these farms were never such intensive cereal producers as those further south. Claims for their 'high state of cultivation', which usually means a high level of grain production, were somewhat exaggerated. On the other hand, the emphasis on cattle accommodation is an indication of their importance. It is perhaps surprising that when so little of the farm (less than a third in any one year) was devoted to cereals, such emphasis was attached to the threshing machine. There were plenty of streams so most farmers used water power; at least one farm had a horse powered system and at Clynelish a steam engine was later installed.

How much of the building activity was the tenants' responsibility and how much it was the landlord's is always difficult to assess. The fact that the plans produced in Loch's book of 1820 are very similar suggests that they were designed from a central estate office, but it is likely that the initial cost and responsibilty for erection was the tenant's. Certainly the leases of the newly-created farms lay great emphasis on what was expected of the tenant. He was 'to have reasonable and proper encouragement for building houses and enclosures suitable to the farm and under a plan approved of by the proprietors of their factors.' The fact that in 1810, the tenant at Culmaily was allowed £1,500 in advance was thought of as unusual. The survival of a book of ameliorations for 1824, shows that most tenants could expect no more than an allowance against rents for improvements undertaken. This allowance might be for all the cost, but often it was only for two-thirds. In many cases rents were nominal for the first seven years, on condition that two thirds of the land was being improved at any one time. As was so often the case in the early years of the nineteenth century, it was the tenant rather than the landlord who was responsible for carrying through the improvements. By the mid 1830s, however, this was changing. Specifications for new work were received in the estate office and approved by the factor who was then responsible for paying for them. This shift from the tenant to the agent as initiator and organizer of improvements can be traced on many estates across Britain at about the same date.

As the new farms were created, the sub-tenants were moved to smallholdings. The organization of this was chaotic. Often only six months' warning was given and their new holdings were not ready for them. Most of the coastal tenants were moved to allotments near Helmsdale or at Portgower, the Garties, Clynelish, Inverbrora and Brora. Here long strips of about two acres (1 hectare) were allotted to them. Very few of these were ever walled off, even though they were cultivated individually, but the fence lines remain. The setting up of a distillery at Clynelish, a harbour at Helmsdale and a coal mine at Brora would, it was hoped, provide employment for the displaced people, but all the land improvement, drainage and walling must have provided an enormous amount of work as well. In the county as a whole the population rose by 217 between 1811 and 1821, but this figure masks the huge local changes. The population in the areas cleared for sheep farms dropped dramatically: in Kildonan by over 1,000, whilst in the parishes with the new settlements such as Clyne, it went up.

The small tenants were expected to put up their own buildings, but to a standard approved by the estate. One- or one-and-a-half-storey cottages of stone with one or two rooms and chimneys in the gable, standing beside a small barn and byre, are still typical of the area. The second storey and dormer windows were often a later addition, perhaps built by the estate as they often incorporare a late nineteenth century datestone and an elaborate Sutherland 'S'.

The sheep farms

In contrast to the improvable lands along the coast, much of the interior of Sutherland was moor and so unsuitable for cultivation, except along the fertile straths where settlement had been concentrated. The improving spirit of the age did not believe that any land need be unproductive, and the possibility of creating large sheep farms in these areas was first suggested in 1799.

82 *The farm buildings at Craikaig with an elaborate porch with a dovecote and the Stafford-Sutherland insignia above the gateway, dated 1829*

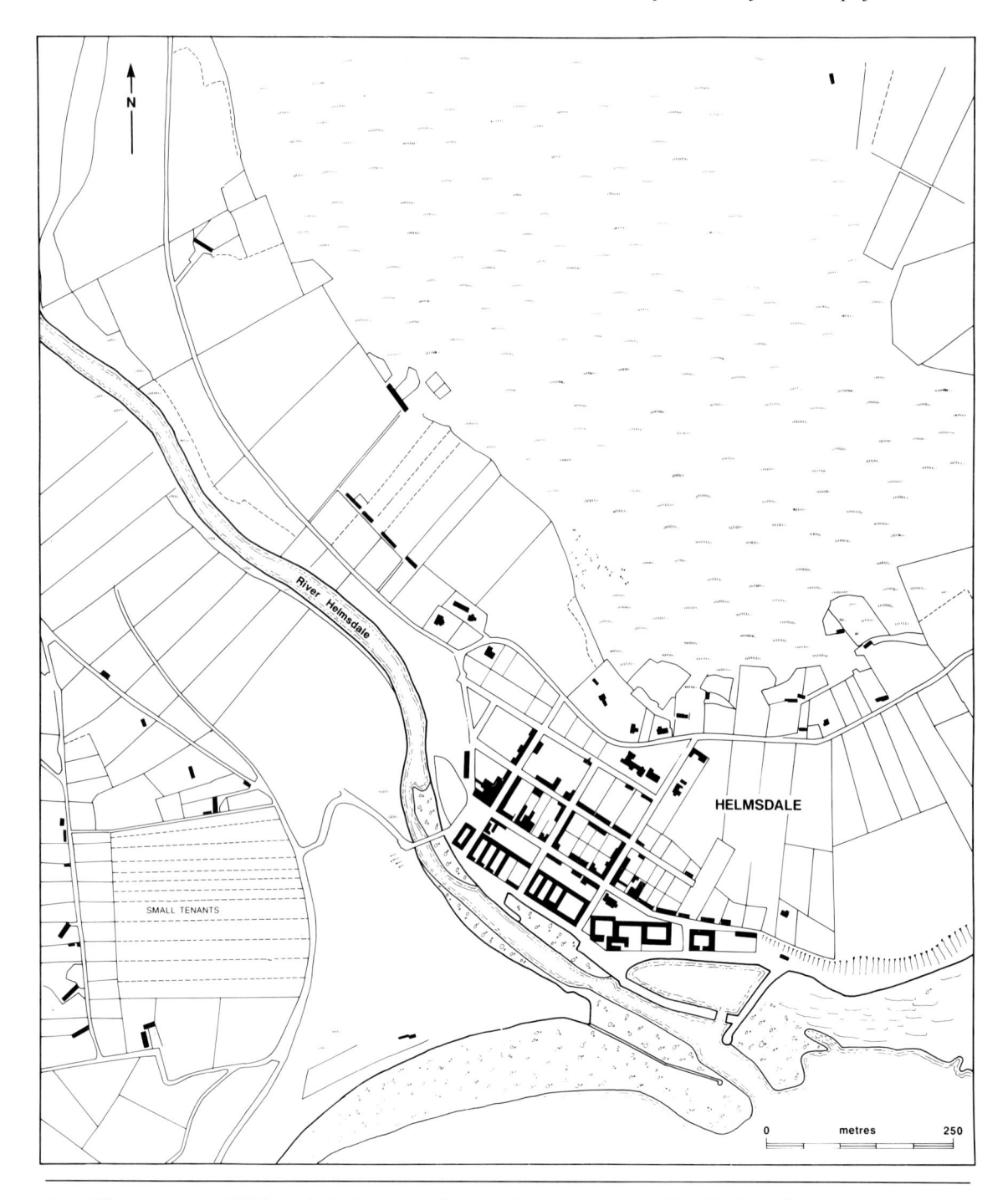

83 *The new town of Helmsdale in the 1880s, showing the narrow strips of land laid out for the small tenants, some with houses at one end (Phillip Judge)*

Strathnaver was one of the areas first considered, and from the beginning the agent realized that there would have to be removals of settlements and 'a considerable thinning' of the population as well as the taking of summer sheilings from the lowland farms. In 1802, two prospective tenants visited the area around Lairg and put their suggestions, in writing, to Earl Gower.

> In the event of our getting the farm, we would immediatly adopt a system perfectly different from that hitherto followed on the estate of Sutherland. In the first instance we would stock wholly with sheep. In the second we would prepare land for commencing turnip husbandry, clear and lay down all the fields in high order with barley and grass seeds, therby improve the land and increase the pasture.

No changes were in fact made until after the securing of the Bridgewater canal wealth in 1805, as the major restructuring that would be necessary would be expensive. In 1806 the agent was instructed to lay out and advertise two sheep farms, one at Shiness and one centred on the parish of Lairg, known as the 'Great Sheep Tenement'. Soon, prospective tenants were visiting the area. The problem with both farms was that they were large and unwieldy and there was little winter feed. Lairg was the first to be let, in 1807, on a nineteen-year lease of £1,300, three times as much as had previously been raised from the area, although the first year's rent was allowed for building sheep fanks (folds), fences and houses. The main problem was that no proper provision had been made for the resettlement of the sixty or so dispossessed small tenants. This confusion was blamed on 'misinformation about the numbers involved' and it was not until 1812–14, when new small farms were created near Lairg church, that the problem was in any way solved.

Evidence for this sheep farm is difficult to come by on the ground; it is not clear where its centre of operations was. By the 1870s the farm was split and two new sets of buildings built, both with extensive cattle accommodation. At Colaboll, on the edge of Loch Shin, there are extensive sheep fanks which may well date from the farm's first creation. On the first edition six inch maps of the 1870s the open moor is dotted with remote and inaccessible shepherd's cottages, with a sheep fold beside them and surrounded by an enclosed piece of improved pasture; these huge upland farms needed little more in the way of buildings.

The second farm, at Shinness, was let in 1808 for £400. It was estimated that it could carry 1,000 sheep. The creation of other sheep farms soon followed, always with the problem that the provisions made for resettling the tenants were inadequate. Torrish and Kilkamkrill (now Borrobol) were created in 1813, resulting in the disappearance of eighty-eight names from the Torrish rental, all of whom were given

84 *The sheep clipping shed with hay barn at one end and loft above, at Torrish sheep farm*

only six months notice to quit. Kilkamkrill was one of the largest sheep farms. Loch wrote in 1820 that 'a set of suitable farm offices had been erected.' These were very different from those of the lowland farms and consisted of large cobbled, stone-walled fanks for sorting and dipping sheep; small low sheds with well-spaced arched openings and flagged floors with wool lofts or feed stores above, known as clipping sheds; and sometimes a small barn or byre for the shepherd. Rarely did the tenant live on his sheep farm as he frequently held a lowland farm as well, where his house would be, so the only houses were small-scale shepherds' houses beside the few necessary buildings. The most controversial sheep farm was that of Rhiloisk, in Upper Strathnaver, also created in 1813 and rented to Patrick Sellar, whose brutal methods of carrying out evictions led to his being tried for murder. Following his acquittal, Sellar remained on his Strathnaver farm during the prosperous mid-century years of high wool prices, making enough money to buy a £20,000 estate in Argyll, to which he finally retired. In spite of his acquittal the estate still came under a great deal of criticism for its methods. In retrospect, the agent agreed that the people had been given too little notice and

> Should have had longer time to move off . . . the fact is that all sheep farmers in making offers should shape them so as to expect little benefit the first summer and our friend Sellar should have known this, but has dearly paid for his rashness.

After an eight year moritorium clearances began again, but the estate was far more cautious, allowing much longer for tenants to resettle. There were, however, still riots and claims of inhumanity. The sheep farms themselves have left less of a mark on the landscape than the coastal farms, but like the coastal farms, they involved the wholesale removal of an earlier way of life and the evidence for it. 15,000 people altogether were moved in the decade up to 1816 and many more, particularly from the west, by 1850. James Loch, who took over total control of the estate after the retirement of Young in 1816, was sure that the clearances were justified and claimed that it was the collapse of the kelp industry, on which the small coastal tenants had depended, and the unreliable herring catch that had caused the distress. In 1852, he was still talking as if the beneficial results of the clearances, which had, after all been going on for nearly half a century, were in the future.

By the 1870s sheep farming was becoming less profitable; some of the farms were being split and cattle steadings built in the better areas. The first lettings for sport had been made in the 1830s but had led to friction between the game keepers and sheep farmers. From the 1870s, the poorer land was being turned over to deer and hunting lodges were being built, often on a grand scale. Game keepers replaced shepherds and venison larders replaced the clipping sheds beside their houses.

Criticism of the system of huge upland sheep farms, with little winter feed or opportunity to grow fodder such as turnips, was increasing by 1820, and the fact that many of them were let with lowland farms shows the impracticality of the system. When the Tongue estate came into Sutherland hands, estate policy, now under the control of James Loch, had changed and the new farms created contained both upland and lowland. The buildings at Ribigil and Melness both have extensive cattle steadings and barns as well as sophisticated sheep handling facilities, including indoor slate-lined sheep dips. A surviving plan of 1833 shows the reorganization of the area into thirty-five sheep farms, involving the subdivision of many of the earlier ones. Loch felt there was a need for a middling group of farms and so favoured the splitting of farms, to provide a more stable society 'to keep up the gradation of ranks, so necessary in all countries, and which perhaps in Sutherland prevails too little.' However, it is not clear how many of these smaller farms were created. There is very little sign of them on the ground.

The question that must be asked is whether the great cost, both in financial terms to the estate and social disruption to the inhabitants, was worth it. The cost in social disruption is impossible to gauge, but feelings still run high locally when the subject of the clearances is raised. Financially, the income of the estate did triple between 1803 and 1817 (a time when rents were rising everywhere), but expenditure had gone up nine times. Between 1812 and 1817, £29,427 more than the estate had produced was spent, whilst in the previous nine years income had exceeded expenditure by nearly £1,900. Much money had certainly been wasted, partly through lack of a general strategy for improvements and the rushed nature of some of the changes. There was plenty for James Loch to criticize when he took over affairs in 1816. His first act was to curtail expenditure until a management plan had been drawn up. However, expenditure continued and about half a million pounds had been spent on the estate, with an equal amount on new purchases, by 1833. In this year, the Duke died which meant his son, Earl Gower, inherited his title and so was no longer eligible for the Bridgewater money; thus this important source of wealth was no longer available. Instead, the Dowager Duchess was to be expected to live off her Sutherland income. This proved impossible and large debts built up. When the Dowager died in 1839, the estate reverted to the new Duke who was constantly put under pressure from Loch to retrench.

A second period of transformation, on a rather smaller scale, took place in the 1870s. At Shinness 1,175 acres (475 hectares) of peat bog were trenched and deep-ploughed using steam power. Four new farms were created with substantial stone-built cattle courts and barns. This extremely expensive scheme was followed by another less ambitious project covering 500 acres (200 hectares) near Kinbrace station. Here tin and wooden buildings, as well as at least two concrete farmsteads, were erected to save the expense of stone. The onset of agricultural depression meant that all work ceased in 1882, and the new arable that produced much-needed winter fodder for the upland sheep only remained in use for a short time.

Whatever mistakes were made, there is no doubt that the landowners and their agents believed in what they were doing. When William Young, the factor who had been responsible for much of the major expenditure, retired in 1816, he wrote, 'My favourite object through life has been to drain moors and improve wastes.' This, he certainly achieved, albeit at great cost, and the landscape of much of Sutherland is testimony of that work.

85 *The elaborate steadings at Clynelish built in 1871 showing, from left to right the separate manure house, the steam engine house attached to the barn and the stables with granary over. The main area of cattle courts is behind the barn*

Epilogue

This book has attempted to trace the development of the farming landscape from the earliest times to the present day and to indicate where this landscape is still visible today. For most of the time that man has been farming in Britain no other evidence for his activities survives. There are no written records for the 5,000 years before the birth of Christ, and for much of the period since then manuscript sources are scanty. It is only in the last 200 years that the majority of farmers have become literate, and so most of our knowledge of the history of farming must be based on the landscape and its buildings. Whilst this has long been accepted for the pre-historic period, which is the traditional stamping ground of the archaeologist, it is just as true for later years. We would know little about the lifestyle and type of farming on the Radnorshire farm of Cilewent, described in Chapter 9, if it were not for the survival of the longhouse there. Not only does the building reveal the farming system at the time it was abandoned, but by piecing together the building's history, it is possible to understand changes both in farming practice since the seventeenth century and in the prosperity of its owners.

Field patterns, too, are evidence for both farming practice and the social structure of a region. Within lowland Britain contrasts are obvious. Whilst many of the small fields across the heavy clays of, for instance south Norfolk and Suffolk, have been swept away, a few pockets survive as an indication of the patchwork of small hedged fields, mostly under 5 acres (2 hectares) that once covered these areas. Where the fields have gone, the narrow windy roads, the irregular mesh of footpaths, and the fact that most of the farms are not concentrated along roads or in villages, but are scattered down tracks, are all indications of an earlier landscape which has proved difficult to erase. Here we have regions dominated by owner-occupiers, where livestock kept in small fields played an important part in the economy. The survival of fine seventeenth-century and earlier farm houses, often within moated sites and with impressive timber-framed barns shows that by the end of the Middle Ages these areas were dominated by a group of prosperous farmers rather than a single landlord. Such landscapes are typical both of the areas of early settlements in the clays of East Anglia and the midland river valleys, and of the secondary, thirteenth century colonization of the old forested areas of Arden, the Weald and the Forest of Dean.

This landscape contrasts with others on lighter soils, where the wide open views are of large rectangular fields, with nineteenth-century farms either along the roads or up straight farm drives. Some older farmsteads survive, but their land has been re-laid out around them. Villages consist of neat rows of labourers' cottages as a large workforce was needed on these extensive farms. These are the landlord-controlled regions, where open commons survived longest and so were ripe for improvement in the nineteenth century. Some parishes contain little more than a single farm and a few cottages. Cereals dominated these areas and whilst grass and roots were grown as break crops, livestock, either cattle, winter-fed in the substantial cattle yards beside the house, or sheep in the turnip fields or on the meadow grasses were seen as the manure producers for intensive grain production. These contrasts can be seen across all regions of lowland Britain, and despite recent hedge removals and changes in farm layout, are still obvious on the ground, although not necessarily clear from the documents. The contrast between these two landscapes is often accentuated by their close proximity to each other: the heavy-clay villages of Suffolk border on the open Brecklands; the late enclosures of the Cotswolds run down to the old fields of the Thames and Severn valleys and their tributuries.

Of equal interest and significance as the difference between these farming patterns is the intermingling of the two types. In parishes such as Cawston in north-east Norfolk there are old farm sites in the village, a relic of the open field days before the local enclosure act of 1801, as well as isolated farmsteads in areas of obviously ancient enclosed fields. Beside these are the post-enclosure farms out in regular fields. A simple distinction between 'ancient' and 'planned' landscapes is obviously not possible here.

As well as the landscape itself, the individual farmsteads, too, have much to tell us. The houses are an indication of the prosperity of the farmers, and their alterations over time point to periods of prosperity and depression. Many a farmhouse was re-fronted in a classical style during the Napoleonic War boom and then either a wing, or at least a bay window, added during the 'high farming' years of the mid-nineteenth century. The swimming pools and conservatories tell of a more recent period of prosperity.

Regional differences are also reflected in buildings. Many upland areas saw their most wealthy times before the cereal boom of the Napoleonic Wars. The seventeenth-century statesman houses of the Lake District reflect the solid, if not flamboyant, lifestyle provided by the breeding of livestock for fattening further south. The enclosure of the uplands produced its rash of Georgian and Victorian residences, from the elegant classicism of the Yorkshire Sledmere estate to the plain farmhouses Exmoor.

The farm buildings, too, are evidence, not only of changing farming practices and the actual needs of the farmer, but also his mood and aspirations. Some of the more extravagant farmsteads of the mid-nineteenth century indicate the optimism of those years and a belief that agriculture played just as important a part in the economy as industry with its often ostentatious factories. Again, it is the alterations to buildings that are as interesting as the original buildings themselves. Only on the great estates, where money was easily available, were farms demolished and a completely new start made.

The gradual accumulation of buildings on the more 'typical' farms of the 'average' farmer helps us piece together a story which is often not told in the documents. There has been much discussion as to how long the 'peasant' continued to exist in English society. He managed to escape comment by the agricultural 'improvers', but sometimes his small farm and its buildings survive.

Agricultural history has, in the past, largely been written from the point of view of the improvers. Historians have tended to accept the views of the agricultural writers of the agricultural revolution and their descriptions of farmers and regions as 'improving' or 'backward', without due regard for the local conditions and the sort of farming suited to each area. As early as the 1870s, the flockmasters on the great sheep farms of Sutherland were realizing that their grazing grounds did not produce the feed there had been fifty years previously. They were suffering from a lack of careful manuring which had been part of the system of farming of the small tenants they had displaced and so much despised.

A recent article has suggested that the traditional Gaelic breast plough or 'caschrom', sneered at as a primitive tool by the improvers from the south, was, in fact well suited to the conditions for which it had been developed.[1] In the same way as our dismissive attitude towards the agricultural systems of third world countries and primitive societies is now being replaced by a respect for some of their more ecologically sound practices, so we are coming to be more cautious in our acceptance of the views of the nineteenth century 'improvers' and their more intensive systems of agriculture. We were, for instance, full of praise for the work of the Duke of Bedford in the fens, but we now know that increased drainage has led to the shrinking of the peat and increasing soil erosion in dry seasons. Perhaps a more 'primitive' grazing system for the area would in the long run have proved more 'progressive'.

An example of the belief that a development in one region could be universally applied to others can be seen in the case of water meadows. This system of winter flooding of riverside meadows to provide an early feed for sheep was developed along the chalk valleys of the southern counties and is described in Chapter 6. It was seen by the agricultural historian, Eric Kerridge,[2] as the basis of agricultural improvement in these areas because it allowed for the keeping of more sheep and thus the production of more manure. Later in the year, when the sheep had moved on to their summer pastures the meadows were mown, producing a heavy crop of hay. Their obvious advantages were noted by the late eighteenth-century agricultural writers and promoters of 'improved' agriculture who could not understand why they had not been adopted elsewhere. It was generally attributed to laziness and ignorance. Some nineteenth-century efforts were made to create water meadows away from their original southern heartland, in the Midlands and East Anglia as well as on the uplands of Exmoor and Scotland. The editor of the *Journal of the Royal Agricultural Society of England*, and enthusiastic experimenter, Philip Pusey, was an advocate of them, but by the 1840s they were mostly the work of aristocratic landlords or larger tenants. Many of these schemes were soon abandoned for a variety of reasons. One was that it was difficult to get enough head of water in the shallow valleys of the lowlands. The leats for the Duke of Portland's Nottinghamshire scheme were over two miles (3 kilometres) long. This could only be done at the tremendous cost of £20 an acre for earth moving and engineering. Fifty acres (20 hectares) would thus cost £1,000: as much as a major farm building improvement. Early grass was not as necessary in areas where turnips were widely grown and, in practice, water meadows tended to be used for hay rather than spring grazing. It is also likely that as the weather was that much colder in the east and in areas above 1,000 feet (300 metres), the system could be frozen up for weeks at a time in the winter. Although there is evidence in some of the East Anglian river valleys for water meadows, and some of the more expensive schemes on the Holkham estate were operating until the end of the nineteenth century, it is clear that the majority were not a success and not suited to the area. It was common sense rather than laziness which had prevented them being built before. In this case the landscape evidence shows us failed improvements which are, in their way, of as much historical significance as successful schemes.[3]

The landscape therefore has much to teach us about the farming past. Its study should not simply be seen as a scene-setting sideline, but as an integral part of agricultural history. Documentation, it should be remembered, is extremely selective. The landscape has taken a tremendous battering over the last thirty years as hedges and woods have been removed, wetlands and pastures ploughed, and buildings altered beyond recognition. Yet it still presents an unbiased view of the past, not weighted towards the great and the good, as the documents tend to be.

It is also true that a greater understanding of the interest and antiquity of our countryside leads to a greater appreciation of its value, and thus a greater chance for its survival in all its fascinating diversity.

Notes to the text

Chapter 1
1. C. Taylor in W.G. Hoskins, 1988, p. 15
2. C. Taylor, 1983, p. 63
3. F. Pryor, 1991, *passim*
4. A. Fleming, 1988, *passim*
5. H.C. Bowen 'Celtic Fields and Ranch Boundaries in Wessex' in S. Limbrey and J.G. Evans (eds), 1978, pp. 115–22
6. O. Rackham, 1986, p. 183
7. J. Nankveris, 1984, *passim*
8. C. Taylor in W.G. Hoskins, 1988, p. 16
9. P. Reynolds, 1979, *passim*
10. O. Rackham, 1986, pp. 158–60
11. A. Davison, 1990, p. 31 and 34
12. T. Williamson, 1986, *passim*
13. C. Taylor in W.G. Hoskins, 1988, p. 16
14. M. Reed, 1990, p. 92
15. R. Hanley, 1987, *passim*
16. S.S. Fryer, 1967, p. 273
17. C. Taylor in W.G. Hoskins, 1988, p. 16
18. B. Dix (ed.), 1987, pp. 16–18
19. C. Taylor, 1983, p. 93
20. G. Foard and J. Pearson, 1985, pp. 5–6 and 1987, pp. 11–12
21. C. Taylor, 1983, p. 96
22. W. Rodwell in C. Bowen and P.J. Fowler (eds), 1978, p. 95 and 92; and in D.G. Buckley, 1980, pp. 59–75
23. C. Taylor, 1983, p. 111
24. *Ibid.* p. 105
25. R.J. Silvester, 1991, *passim*

Chapter 2
1. C. Taylor in W.G. Hoskins, 1988, pp. 41–2
2. M. Aston, in R. Higham (ed.), 1989, pp. 24–5
3. G. Foard and T. Pearson, 1985, p. 5 and B. Dix (ed.), 1987, pp. 11–16
4. O. Rackham, 1986, pp. 79–80
5. P.J. Crabtree, 1990, *passim*
6. G. Foard, J. Pearson, 1985, pp. 6–7, B. Dix (ed.), 1987, p. 18
7. B. Dix (ed.), 1987, p. 20
8. Beresford and J. Hurst, 1990, pp. 95–6
9. M. Aston in Higham (ed.), 1989, p. 28
10. D. Yaxley, 1980, vol. 2, p. 578
11. M. Harvey, 1983
12. C. Dyer, 1991, pp. 60–2.
13. O. Rackham, 1986, p. 84

Chapter 3
1. C. Taylor in W.G. Hoskins, 1988, pp. 41–2
2. T. Williamson and L. Bellamy, 1987, p. 38
3. M. Harvey, 1983
4. M. Beresford and J. Hurst, 1990, p. 96
5. D. Hall, 1980, *passim*
6. C. Taylor, 1975, p. 83
7. *Ibid.* p. 80
8. J.V. Becket, 1989, *passim*
9. J.S. Moore, 1965, *passim*
10. O. Rackham, 1986, p. 97
11. A. Grant in G. Asthill and A. Grant (eds), 1988, pp. 150–7

Chapter 4
1. S. Everett, 1968, pp. 54–60

2 J. Grigg in Asthill and A. Grant (eds), 1988, p. 108
3 A. Grant in G. Asthill and A. Grant (eds), 1988, p. 151
4 F.G. Davenport, 1967, p. 21
5 E. Wiliam, 1986, p. 4
6 M. Reed, 1990, p. 138
7 R.J. Brien, 1989, pp. 12–14
8 C. Taylor, 1973, p. 105
9 C. Platt, 1969, *passim*
10 S. Denyer, 1991, p. 78
11 R. A. Donkin, 1963, pp. 181–98
12 S. Denyer, 1991, pp. 76–7
13 J. Bond, 1979, pp. 59–75
14 J. Bond, 1973, pp. 1–61
15 Guide book to Cressing Temple (Essex County Council).
16 W. Horn and E. Born, 1965, *passim*
17 J. Lake, 1989, p. 62
18 N. Alcock, 1981, *passim*
19 R.J. Silvester, 1988, map insert
20 A. Fleming and N. Ralph, 1982, pp. 101–137
21 S. Denyer, 1991, p. 111
22 C. Dyer, 1990, pp. 97–121
23 *Ibid.*
24 J. Sheppard, 1976, pp. 3–20
25 C. Dyer, 1991, p. 29
26 J.S. Moore, 1965, *passim*
27 J. Hurst in Camb. Ag. Hist. vol. 2, 1988, pp. 861–915

Chapter 5

1 F.G. Davenport, 1967, *passim*
2 C. Dyer, 1991, p. 52
3 C. Taylor, 1974, p. 133
4 A. Fleming and N. Ralph, 1982, pp. 101–137
5 G. Beresford, 1979, pp. 98–158
6 P. Wade Martins, vol. 10, 1980, pp. 93–161
7 C. Dyer, 1991, pp. 54–5
8 M. Aston in R. Higham (ed.), 1989, p. 36
9 J. Ravensdale, 1974, *passim*
10 M. Aston, 1983, pp. 71–104
11 F. Emery, 1974, p. 107
12 J. Bond, 1979, *passim*
13 D. Clarke, 1972, pp. 25–7
14 N.W. Alcock, 1981, *passim*
15 J. Thirsk, 1987, p. 61

Chapter 6

1 F. Emery, 1974, p. 101
2 M. Reed, 1981, pp. 60–8
3 S. Wordie, 1983, pp. 483–505
4 W. Marshall, 1787, vol. 1. p. 4
5 S. Wordie, 1983, pp. 483–505
6 O. Rackham, 1986, p. 188
7 *Ibid.* p. 179
8 R.C. Allen, 1992, p. 128
9 P. Glennie in R. Glasscock (ed.), 1992, p. 135
10 J. Bettey, 1977, pp. 37–43, G.G. Bowie, 1987, pp. 151–8, E. Kerridge, 1967, pp. 251–67
11 H.C. Derby, 1983, *passim*
12 P. Beacham (ed.), 1990, pp. 47–60
13 S. Denyer, 1991, *passim*
14 E. Wiliam, 1986, p. 5
15 P. Smith in L.I. Foster and L. Alcock, 1963, p. 435

Chapter 7

1 A. Young, 1808, p. 25
2 Walpole archives. Wolterton, Norfolk MS leases 4/1–4
3 R.A.C. Parker, 1975, pp. 84–8
4 Heydon, 1780s maps Norfolk Record Office MFRO 334/1, 3.
5 A. Young, 1808, p. 210
6 O. Rackham, 1986, p. 190
7 M. Turner, 1984, pp. 66–7
8 T. Stone, 1794, p. 143
9 M. Havinden, 1981, p. 203
10 S. Wade Martins, 1980, p. 88
11 M. Williams, 1970, *passim*
12 Arthur Young, 1808, p. 61
13 *Ibid.* p. 234
14 *Ibid.* p. 32
15 M. Turner, 1984, pp. 64–73
16 S. Wade Martins, 1991, pp. 143–4
17 M. Havinden, 1981, pp. 203–24
18 W. Pitt, 1788, p. 400

19 J.E.C. Peters, 1969, pp. 113–15
20 S. Wade Martins, 1991, p. 199
21 R.W. Brunskill, 1982, pp. 113–15
22 J.M. Robinson, 1983, *passim*
23 S. Wade Martins, 1980, pp. 48–9
24 J. Popham, 1986, pp. 128–32

Chapter 8
1 E. Wiliam, 1986, p. 5
2 G.R.J. Jones in A.R.H. Baker and R.A. Butlin, pp. 430–9
3 E. Wiliam, 1986, p. 8
4 *Ibid*. p. 4
5 *Ibid*. p. 14
6 *Ibid*. p. 50
7 E. Wiliam, 1992, *passim*
8 R.J. Brien, 1989, *passim*
9 Lecture given by G. Stell, RCAHMS, at the Historic Farm Buildings Group's autumn conference, Edinburgh, 1991
10 R.J. Brien, 1989, p. 18
11 J.M. Robinson, 1983, p. 18
12 J. Caird in M.L. Parry and T.R. Slater, 1980 pp. 203–22
13 R.J. Brien, 1989, p. 21
14 A. Fenton and B. Walker, 1981, *passim*
15 S. Wade Martins, 1987, pp. 23–32

Chapter 9
1 R. Brigden, 1986, pp. 203–8
2 S. Wade Martins, 1980, p. 121
3 A.D.M. Phillips, 1989, *passim*
4 S. Wade Martins, 1980, p. 100
5 R.N. Bacon, 1844, pp. 379–85
6 C.S. Orwin and E.H. Wretham, 1971, p. 196
7 R. Brigden, 1986, p. 23
8 S. Macdonald in G.E. Mingay, 1981, p. 214
9 D. Spring, 1963, p. 117
10 S. Wade Martins, 1980, pp. 169–72
11 J.B. Denton, 1863, pp. 57–9
12 S. Wade Martins, 1980, p. 119
13 *Ibid*. p. 116
14 P. Pusey, 1851, p. 35
15 C.S. Read, 1858, pp. 265–311
16 S. Wade Martins, 1980, pp. 165–70

17 R. Brigden, 1986, pp. 61–2
18 A. Phillips in Holderness and Turner, 1991, pp. 191–210
19 R. Brigden, 1986, pp. 190–2

Chapter 10
1 H. Bidell, 1907, pp. 385–402
2 J. Brown, 1987, pp. 33–6
3 W.J. Moscrop, 1890, p. 473
4 R. Brigden, 1986, pp. 228–9
5 J. Brown, 1987, p. 50
6 M.A. Havinden, 1966, *passim*
7 S. Wade Martins, 1980, p. 182
8 J.R. Gray, 1971, pp. 171–85
9 H. Williamson, 1986, p. 37
10 J. Brown, 1987, p. 99
11 R. Brigden, 1991, pp. 18–31
12 R. Brigden, 1992, pp. 35–48
13 J. Brown, 1897, pp. 125–46
14 R. Westamott and T. Worthington, 1974, pp. 9–14
15 *Ibid*. pp. 15–18
16 *Ibid*. pp. 19–22

Chapter 11
1 J. Popham, 1986, pp. 128–32
2 G.E. Fussell, 1948, pp. 296–310
3 C.S. Orwin and R.J. Sellick, 1970, *passim*

Chapter 12
1 H.C. Derby, 1956, 1983
   C. Taylor, 1973
2 Bedford, Duke of, 1897
3 H.C. Darby, 1983
4 C. Taylor, 1973

Chapter 13
1 J. Loch, 1820
2 E. Richards, 1973
3 R.J. Adam, 1972
4 H. Fairhurst, 1964, 1987

Epilogue
1 R.A. Dodgshon, 1992
2 E. Kerridge, 1967, pp. 251–67
3 S. Wade Martins and T. Williamson, 1994

Bibliography

Adam, R.J. (ed.), *Papers on Sutherland Estate Management 1802–1816*, 2 vols, Edinburgh, 1972

Alcock, N.W. *Cruck Construction*, CBA Research Report 42, London, 1981

Allen, R.C. *Enclosure and the Yeoman Farmer*, Oxford, 1992

Asthill, G. and Grant, A. (eds), *The Countryside of Medieval England*, Oxford, 1988

Aston, M. *Interpreting the Landscape*, London, 1985

'Deserted Farmsteads on Exmoor', *Proceedings of the Somerset Archaeological and Natural History Society*, 127 (1983), pp. 71–104

Aston, M., Austin, D. and Dyer, C. (eds), *The Rural Settlements of Medieval England*, Oxford, 1989

Bacon, R.N. *Agriculture of Norfolk*, London, 1844

Baker, A.R.H. and Butlin, R.A. (eds), *Studies of Field Systems in the British Isles*, Cambridge, 1973

Beacham, P. (ed.), *Devon Building*, Exeter, 1990

Beckett, J.V. *The Agricultural Revolution*, Oxford, 1990

Laxton, England's last Open Field Village, Laxton, 1989

Bedford, Duke of, *The Story of a Great Estate*, London, 1897

Beresford, G. 'Three Medieval Deserted Villages on Dartmoor', *Medieval Archaeology*, 23 (1979), pp. 98–158

Beresford, M. and Hurst, J. *Deserted Medieval Villages*, London, 1971

Wharram Percy, London, 1990

Bettey, J. 'The development of water meadows in Dorset in the 17th century', *Agricultural History Review*, 25 (1977), pp. 37–43

Bidell, H. 'Agriculture', *Victorian County History*, Suffolk, London, 1907, pp. 385–402

Bond, J.C. 'The estates of Evesham Abbey: a preliminary survey of their medieval topography', *Vale of Evesham Historical Society Papers*, 4 (1973), pp. 1–61

'The reconstruction of the medieval landscape. The estates of Abingdon Abbey', *Landscape History*, 1 (1979), pp. 59–75

Bowen, C. and Fowler, P.J. (eds), *Early Land Allotment*, Oxford, 1978

Bowie, G. 'Water meadows in Wessex; a re-evaluation for the period 1640–1850', *Agricultural History Review*, 35 (1987), pp. 151–8

Brigden, R. *Victorian Farms*, London, 1986

'Richard Borlase Matthews and his Great Felcourt "Electro Farm"', *Journal of The Historic Farm Building Group*, 5 (1991), pp. 18–31

'Bucking the Trend: New farms between the Wars', *Journal of The Historic Farm Buildings Group*, 6 (1992), pp. 35–48

Brown, J. *Agriculture in England, 1870–1947*, Manchester, 1987

Brunskill, R.W. *Traditional Farm Buildings of Britain*, London, 1982

Buckley, D.G. (ed.), *Archaeology in Essex to AD 1500*, CBA Research Report 34, London, 1980

Cadman, G. *Raunds 1981*, Nothampton, 1981

Cadman, G. et al. *Raunds–Furnells*, Northampton, 1983

Cantor, L. *The Changing English Countryside 1400–1700*, London, 1987

Cantor L. (ed.), *The English Medieval Landscape*, London, 1982

Chambers, J.D. *Laxton*, London, 1964

Chambers, J.D. and Mingay, G.E. *The Agricultural Revolution*, London, 1966

Clapperton, C.M. (ed.), *Scotland a new study*, Newton Abbott, 1983

Clarke, D. 'Penine aisled barns', *Vernacular Architecture*, 4 (1972), pp. 25–7

Crabtree, P.J. *West Stow, Suffolk: Early Anglo-Saxon Animal Husbandry*, Ipswich, 1990

Davenport, F.G. *The Economic Development of a Norfolk Manor 1086–1565*, reprint, 1967

Davison, A. 'The evolution of settlement in three parishes in south-east Norfolk' *East Anglian Archaeology*, 49, Gressenhall, 1990

Denton, J.B. *The Farm Homesteads of England*, London, 1863

Denyer, S. *Traditional Buildings and Life in The Lake District*, London, 1991

Derby, H.C. *The Draining of the Fens*, Cambridge, 1956

The Changing Fenland, Cambridge, 1983

Dix, B. (ed.), *Raunds Area Project 2nd interim report*, Northampton, 1987

Donkin, R.A. 'The Cistercian Order in Medieval England', *Transactions of the Institute of British Geographers*, 33 (1963), pp. 181–98

Dyer, D. *Lords and Peasants in a Changing Society*, Cambridge, 1980

Hanbury: Settlement and Society in a Woodland Landscape, Leicester, 1991

'Dispersed settlement in early medieval England; a case study of Pendock, Worcestershire', *Medieval Archaeology*, 34 (1990), pp. 97–121

Emery, F. *The Oxfordshire Landscape*, London, 1974

Everett, S. 'The Domesday geography of three Exmoor parishes', *Proceedings of the Somerset Archaeological and Natural History Society*, 112 (1968), pp. 54–60

Everitt, A.M. *Landscape and Community in England*, London, 1985

Fairhurst, H. 'The Survey for the Sutherland Clearances', *Scottish Studies*, 8 (1964), pp. 1–18

'Rosal, A deserted township in Strathnaver, Sutherland', *Proceedings of the Society of Antiquaries of Scotland*, 100 (1987–8), pp. 135–169

Fenton, A. and Walker, B. *The Rural Architecture of Scotland*, Edinburgh, 1981

Fleming, A. *The Dartmoor Reaves. Investigating Prehistoric Land Divisions*, London, 1988

Fleming, A. and Ralph, N. 'Medieval settlement and landuse on Holme Moor, Dartmoor: the landscape evidence', *Medieval Archaeology*, 26 (1982), pp. 101–37

Foard, G. and Pearson, T. *The Raunds Area Project: 1st interim Report*, Norhampton, 1985

Frere, S.S. *Britannia*, 3rd ed., London, 1987

Frere, S.S. and St Joseph, J.K.S. *Roman Britain from the Air*, London, 1983

Foster, I.I. and Alcock, L. (eds), *Culture and Environment*, London, 1963

Fox, H.S.A. and Butlin, R.A. *Change in the Countryside*, London, 1979

Fussell, G.E. '"High Farming" in the North of England 1840–1880' *Economic Geography*, 24 part 4 (1948), pp. 296–310

Glasscock, R. *Historic Landscapes of Britain from the Air*, Cambridge, 1992

Fussell, G.E. 'The Reclamation of the Wolds', *Journal of the Land Agents' Society*, 53 (April 1954), pp. 159–61

Gray, J.R. 'An industrial farm estate in Berkshire', *Industrial Archaeology*, 8 (1971), pp. 171–85

Hall, D. *Medieval Fields*, Aylesbury, 1980

Hanby, R. *Villages in Roman Britain*, Aylesbury, 1987

Harvey, M. 'Planned Field Systems in East Yorkshire', *Agricultural History Review*, 31 (1983), pp. 91–103

Harvey, N. *Fields, Hedges and Ditches*, Aylesbury, 1976

A History of Farm Buildings in England and Wales, 2nd edn, Newton Abbot, 1984

Havinden, M. *Estate Villages, a study of the*

Bibliography

Berkshire villages of Ardington and Lockinge, Reading, 1966

The Somerset Landscape, London, 1981

Higham, R. (ed.), *Landscape and Townscape in the South-west*, Exeter, 1989

Holderness, B.A. and Turner, M. (eds), *Land, Labour and Agriculture*, London, 1991

Hooke, D. (ed.), *The Medieval Village*, Oxford, 1985

Horn, W. and Born, E. *The barns of the Abbey of Beaulieu and its Granges*, California, 1965

Hoskins, W.G. *The Making of the English Landscape*, (edition with introduction by C.C. Taylor), London, 1988

Hurst, J. *Cambridge Agrarian History*, vol. 2, pp. 861–915, Cambridge, 1988

Kerridge, E. *The Agricultural Revolution*, London, 1967

The Open Fields, Manchester, 1992

Lake, J. *Historic Farm Buildings*, Blandford, 1989

Loch, J. *An Account of the Improvements on the Estates of the Marquis of Stafford*, London, 1820

Maby, R. *The Common Ground*, London, 1980

Marshall, W. *Rural economy of Norfolk*, 2 vols, London, 1787

Millman, R.N. *The Making of the Scottish Landscape*, Edinburgh, 1975

Mingay, G.E. *The Victorian Countryside*, in 2 vols, London, 1981

Moore, J.S. *Laughton A study in the evolution of the Wealden landscape*, Leicester, 1965

Moscrop, W.J. 'Covered cattle yards', *Journal of the Royal Agricultural Society of England*, 3rd Series, 1 (1890), pp. 473–90

Muir, R. *The Shell Guide to Reading the Landscape*, London, 1981

Nankvernis, J. *The Traditional Farm, Wicca, Zennor, St Ives*, MAFF, 1989

Newby, H. *The Countryside in Question*, London, 1988

Orwin, C.S. and Sellick, R.J. *The Reclamation of Exmoor Forest*, Newton Abbot, 1970

Orwin, C.S. and Wretham, E.H. *History of British Agriculture 1846–1914*, 2nd ed., Newton Abbot, 1971

Parker, R.A.C. *Coke of Norfolk, A financial and agricultural study 1707–1842*, Oxford, 1975

Parry, M.L. and Slater, T.R. *The making of the Scottish Countryside*, London, 1980

Peters, J.E.C. *The development of Farm Buildings in West Lowland Staffordshire up to 1880*, Manchester, 1969

Phillips, A.D.M. *The Underdraining of Farmland in England during the nineteenth century*, Cambridge, 1989

Pitt, W. 'Buildings of a farm', *Annals of Agriculture*, 9 (1788), p. 400

Platt, C. *The Monastic Grange in Medieval England*, London, 1969

Popham, J. 'Sir Christopher Sykes at Sledmere', *Country Life*, 16 January 1986, pp. 128–32

Pryor, F. *Fengate*, London, 1991

Pusey, P. *What Ought Landlords and Farmers To Do?* London, 1852

Rackham, O. *The History of the Countryside*, London, 1986

Ravensdale, J. *Liable to flood*, Cambridge, 1974

Read, C.S. 'Recent improvements in Norfolk farming' *Journal of the Royal Agricultural Society of England*, 19 (1858), pp. 265–311

Reed, M. (ed.), *Discovering Past Landscapes*, London, 1984

The Landscape of Britain, London, 1990

'Pre-Parliamentary Enclosure in the East Midlands', *Landscape History*, 3 (1981), pp. 60–8

Reynolds, P. *Iron Age farming, The Butser Experiment*, London, 1979

Richards, E. *The Leviathan of Wealth*, London, 1973

Robinson, J.M. *Georgian Model Farms*, Oxford, 1983

Rowley, T. *Villages in the Landscape*, London, 1978

The Origins of open-field agriculture, London, 1981

Sawyer, P. *Introduction to Medieval Settlement*, London, 1976

Sheppard, J. 'Medieval Village plans in northern England', *Journal of Historical Geography*, 2 (1976), pp. 3–30

Shoard, M. *The Theft of the Countryside*, London, 1980

Silvester, R.J. *The Fenland Project, No. 3: Norfolk Survey*, Marshland and the Nar Valley, Gressenhall, 1988
  Fenland project (4) Norfolk Survey, The Wissey Embayment and Fen Causeway, Gressenhall, 1991
Spring, D. *The English Landed Estate in the nineteenth century*, Baltimore, 1963
Stone, J.F.S. *Wessex*, London, 1958
Stone, T. *General View of the Archaeology of Lincolnshire*, London, 1794
Taylor, C. *The Cambridgeshire Landscape*, London, 1973
  Fields in the English Landscape, London, 1975
  Village and Landscape, London, 1983
Thirsk, J. *Agrarian regions and agrarian history in England 1500–1750*, London, 1987
  (ed.) *The Agrarian History of England and Wales V, 1640–1750*, Cambridge, 1985
  English Peasant Farming, London, 1957
Turner, M. *Enclosures in Britain, 1750–1830*, London, 1984
Wade-Martins, P. 'Village sites in Launditch hundred', *East Anglian Archeology*, 10, Gressenhall, 1980
Wade Martins, S. *A Great Estate at Work, Holkham and its inhabitants in the nineteenth century*, Cambridge, 1980
  Historic Farm Buildings, London, 1991
  Eigg, an Island Landscape, Stirling, 1987
Wade Martins, S. and Williamson, T. 'Floated water meadows in Norfolk – a misplaced innovation', *Agricultural History Review*, 42, part 2, 1994
Welsh, M. *Anglo Saxon England*, London, 1992

Westmacott, R. and Worthington, T. *New Agricultural Landscapes*, Cheltenham, 1974
Whittington, G. and Whyte, I.O. (eds), *The Historical Geography of Scotland*, Edinburgh, 1983
Wiliam, E. *Traditional Farm Buildings in north-east Wales*, Cardiff, 1982
  Historic Farm Buildings of Wales, Edinburgh, 1986
  Welsh Long-houses Four centuries of farming at Cilewent, Cardiff, 1992
Williams, M. *The Draining of the Somerset Levels*, London, 1970
Williamson, H. *The Story of a Norfolk Farm*, London 1941, reprint 1986
Williamson, T. 'Explaining Regional Landscapes; Woodland and Champion in Southern and Eastern England', *Landscape History*, 10 (1988), pp. 5–14
  'Early co-axial field systems on the East Anglian Boulder Clays', *Proceedings of the Prehistoric Society*, 53 (1986), pp. 419–31
Williamson, T. and Bellamy, L. *Property and Landscape*, London, 1987
Wordie, S. 'Chronology of English Enclosure 1500–1914, *Economic History Review*, 36 (1983), pp. 483–505
Yaxley, D. 'Documentary evidence for North Elmham' in P. Wade-Martins, 'Excavations in North Elmham park', *East Anglian Archaeology*, vol. 9, Gressenhall, 1980
Yelling, J.A. *Common Field and enclosure in England 1450–1850*, London, 1977
Young, A. *General Report on Enclosures*, 1808, reprint New York, 1971

Index

(Sites and information in the captions are indexed as if part of the text.)

Aberdeen Angus (cattle) 96
'Agricultural Revolution' 75, 78, 86, 90, 93, 101, 121
Airde, The, Loch Shin, Sutherland 80
Aisled barns 44–7, 51, 59
Allendale, Northumberland 74
Anglo-Saxon 22, 29, 40
Anglo-Saxon charters 29
Apples 17
Appleton-le-Moors, Yorkshire 31
Ard 9, 14
Ardington, Berkshire 115
Arden, Forest of, Warwickshire 40, 51, 146
Argyll, Duke of 95
Armstrong, Lord 115
Assarts 42, 49, 50, 54–5
Ayrshire (cows) 96

Bacon, R.N. 104
Bagley, Somerset 22
Badminton, Gloucestershire 88
Bald, William, surveyor 98
Bank barns 72–3, 86, 112
Bardolf family 30
Barley 13, 24, 32, 41, 82, 95, 140
Barns 8, 44–7, 51, 52, 60, 69, 70, 71, 78, 82, 83, 85, 88–90, 92, 108, 114, 131, 136, 138, 139
Bastel houses 72, 74
Battery housing (poultry) 117
Beans 13, 32, 41, 140
Beaulieu Abbey estates (Hampshire & Oxfordshire) 46
Bedford, Dukes of 87, 122, 124–132
Bedford, Francis, 4th Earl 68
Bedford river (Cambridgeshire) 68, 69
Bedfordshire 13, 87
Belhaven, Lord 95
Benedictine monks 44
Bentley Grange (Yorkshire) 35
Bere 41, 95, 135, 140
Berkshire 63
Berwickshire 87
Bibury, Gloucestershire 53
Black-faced sheep 95
Black Death 54, 56–9
Blackmoor Forest, Dorset 50
Blairdrummond Moss, Stirling 23
Board of Agriculture 95

Board of Management, Annexed Estates 95
Board of Trustees for Fisheries and Manufacture 95
Bordars 40
Borrobol, Sutherland 143
Boston, Lincolnshire 41
Bourne valley, Dorset 12
Braunton, Devon 35
Brecklands, Norfolk and Suffolk 54, 82, 101, 147
Brecon Forest, Powys 92
Brewhouse 46
Bridges 136
Bridgewater, Duke of 133
Bristol Channel 19, 80
Britford, Wiltshire 67
Brookend, Oxfordshire 57, 63
Brora, Sutherland 141
Buckinghamshire 63
Bullocks, see Cattle
Burnham Westgate, Norfolk 75
Buscot Park, Berkshire 116
Butterwick, Yorkshire 27
Butser Hill, Hampshire 13
Byre 51, 72, 85, 92–4, 112, 136, 138
Byre house, see Longhouse

Caird, James, agricultural writer 106
Caledonian forest 9
Caldy Island, West Glamorgan 80
Cambridgeshire 54, 78, 118
Canna, Inner Hebrides 97
Canterbury, Kent 43
Car Dyke, Lincolnshire 19
Carrots 17
Carse of Gowrie, Perthshire 23
Carthouse 51
Cary river, Somerset 80
Caschrom 148
Castle Acre, Norfolk 76
Castle Carlton, Lincolnshire 30
Cattle 9, 12, 13, 18, 23, 39, 41, 42, 44, 46, 66, 72, 85, 89, 91–3, 96, 97, 106, 107, 109, 140
Cattle boxes 109; see also Loose boxes
Cattle house 41, 92
Cattle sheds 96, 108, 144, 147
Cattle yard 90, 109, 131; see also Covered yards

Cawston, Norfolk 147
'Celtic' fields 16
Cerne Abbas, Dorset 30
Channonz Hall, Norfolk 70
Charlton, Hampshire 13, 19
Chat Moss, Lancashire 122
Chedworth, Gloucestershire 16
Cheshire 63
Cherries 17
Chester, Bishop of 19
Chet river, Norfolk 15
Chickens 14, 46, 59
Chippenham, Cambridgeshire 57
Chollerton, Northumberland 88
Cider mill 46
Cilewent, Radnorshire 92–3, 146
Cirencester Agricultural College, Gloucestershire 101
Cistercian monks 43–4, 46
Civil War (1640–1649) 62, 68, 70
Clanranald, Reginald, clan chief 97
Claying 66, 79, 96
Clearances 96, 122, 135, 144–5
Climate 41, 53
Clipping sheds 143–4
Clover 66, 140
Clover Top Farm, Hertfordshire 117
Clynelish, Sutherland 140, 141
Cobbett, William, writer 87, 128
Cockburn John of Ormiston, landowner, East Lothian 95
Coggeshall, Essex 46
Coke, Thomas William of Holkham, landowner, Norfolk 88
Coleseed 127
Commission for Highland Roads and Bridges 137
Common field, see Open fields
Commons 7, 32, 39, 40, 50, 51, 64, 76, 81, 83, 91, 92, 147
Commonwealth (1649–1660) 66
Concrete 112, 115, 117, 118, 145
Coppicing 23
Corn Laws 101, 103
Corn tax 17
Corpusty, Norfolk 76
Corrugated iron 112, 114, 116
Cornwall 36, 38, 41, 62, 72, 118
Cotswolds, Oxfordshire and Gloucestershire 50, 61, 64, 147

Cottages 96, 116, 129, 146
Coul, Sutherland 136
Courtyard farm plans 86, 88, 96, 136, 138–9
Covered cattle yards 109–12, 114, 139
Cows 92, 98
Cowshed 8
Cragside, Northumberland 115
Craigton, Stirlingshire 94
Craigyload, Perthshire 94
Craikaig, Sutherland 141
Cranbourne, Dorset 14
Crave, Yorkshire 18
Cressing Temple, Essex 44, 45
Cressingham Manor, Norfolk 70
Cricket St Thomas, Somerset 64
Crimplesham, Norfolk 108, 109
Crofts 97, 98, 100
Crosby Ravensworth, Cumbria 18
Crow Hill, Northamptonshire 23
Crowland Abbey, Cambridgeshire 43, 68
Cruck construction 47, 52, 59, 60, 136
Culbone, Somerset 40–1
Culham, Oxfordshire 44
Culmaily, Sutherland 138, 140
Cumbria 17, 18, 103
Cyfeilig, Powys 91

Dairies, dairying, dairy farming 41, 46, 62, 92, 115, 117
Danelaw 27
Dartmoor, Devon 11–12, 21, 47, 54, 72
Dean, G.A., agricultural engineer 107, 109
Deer 7, 13, 100, 104, 121
Deforestation 9, 68
Demesne farming 41, 42, 47, 53, 57, 91
Denton, J.B., agricultural engineer and writer 104, 124, 130
Denver, Norfolk 19
Derby, Earl of 103
Derbyshire 63
Deserted villages 51, 54, 63, 72
Devon 36, 63, 70, 72
Dexter cattle 14
Dickleborough, Norfolk 15
Ditchley, Oxfordshire 18
'Diversification' 119
Domesday book 8, 27–28, 30, 40
Donnington Health, Leicestershire 14
Dorset 64, 67, 118, 120
Dovecotes 44, 46, 51, 96, 112
Drainage 34, 43, 68, 80, 95, 96, 124–8, 132, 145; underdrainage 66, 102–4, 129, 131, 132, 137
Drain pipes 103, 129
Droveways 9, 10, 47
Dunkery, Somerset 22
Durham, Bishops of 30
Durham 37, 87, 103
Dutch barns 114

Eastfield Farm, Lothian 111
Eastwood Manor Farm, East Harptree, Avon 109, 110
Eigg, Isle of, Inner Hebrides 97–100
Electricity 117
Ely, Cambridgeshire 43
Emner 13, 14, 24
Enclosure (*see also* Parliamentary enclosure) 7, 32, 56–8, 61, 62, 64, 67, 68, 72, 75, 76, 78–9, 80–1, 83, 90–3, 96, 121

Estates 7, 8, 19, 30, 41, 66, 75, 78, 87, 92, 105, 116, 121, 122, 147
Essex 15, 63, 69, 114
European Economic Community 118
Evesham Abbey, Gloucestershire 43
Exmoor, Somerset and Devon 22, 41, 56, 121, 148
Eynsham Abbey, Oxfordshire 57

Fallow 32, 89, 113, 131, 135, 140
Famine 42, 135
Farm amalgamations 57, 92, 118, 121
Farm houses 69–72, 83, 90, 92, 96, 98, 129, 132, 139, 147
Farm layout 85, 86, 90; *see also* Courtyard farm plans
Farmstead
  Bronze Age 12
  Iron Age 13–14
  Roman 17, 21
  Medieval 40, 44, 58
  Post-Medieval 83, 129, 147
Fat hen 13
Felsham, Suffolk 38
Fengate, Cambridgeshire 9
Fens, The, Cambridgeshire, Lincolnshire, Norfolk 16, 23, 54, 56, 68–9, 115, 118, 121, 122
Fermtoun 93, 95, 96
Fertilisers 101, 107, 118
Fertility 9, 53–4
Fields
  Bronze Age 7–12
  Iron Age 14–15
  Roman 16, 19
  Medieval 8, 26–7, 30; *see also* Infield-outfield, Open fields, Strip fields
  Post-medieval, *see* Enclosure
First World War 116, 122
Firth of Forth 36
Flag Fen, Cambridgeshire 11, 19
Flanders Moss, Perthshire 23
Flax 12, 24, 32, 41, 140
Flooding 54, 126
Foldcourse system 58
Forest 9, 50, 52, 62
Forncett, Norfolk 41, 42, 53
Foss Dyke, Lincolnshire 19
Fountains Abbey, Yorkshire 43
Frocester, Gloucestershire 47
Frome river, Wiltshire 68
Fulmodestone, Norfolk 76
Furlong 7, 32, 34, 35, 50, 52, 57, 58, 61

Geese 46
Gentry 57
Glastonbury Abbey, Somerset 43, 47, 54
Glastonbury Lake Village 14
Gower Peninsula, West Glamorgan 80
Grain prices 62, 63, 66, 101, 106, 113, 116
Great Coxwell barn, Oxfordshire 46
Great Felcourt, Surrey 117
Great Langdale, Cumbria 49
Greens 50
Green villages 52
Goats 9, 13, 23
Goose-house 41
Gower, Earl 135, 143, 145
Graham, Sir John, landowner, Northumberland 121
Granaries 8, 16, 41, 44, 46, 51, 70, 71, 78, 83, 85, 138
Granges 43, 44, 46

Grassington, Yorkshire 16
Greencaig, Fife 16
Greens 30, 50
Grenstein, Norfolk 54, 55
Grey John of Dilston, landowner, Northumberland 121
Gunton estate, Norfolk 115
Gwithian, Cornwall 35

Haddington, 6th Earl 95
Hafods 91
Hafod Elwy, Denbighshire 42
Hamlets 17, 27, 42, 51, 56, 91, 98
Hampshire 67, 115
Hanbury, Worcestershire 50, 53, 55
Hartland, Devon 56
Hastings, John, farmer, Norfolk 79, 102
Hatton, Warwickshire 63
Havering, Essex 55
Hay 66, 72, 140, 148
Hay dryer 117
Hay-house 41
Hay loft 72, 93
Hebrides 97
Hedges 12, 15, 21, 33, 38, 39, 50, 57, 61, 64, 67, 78, 79, 82, 83, 90, 113, 117, 122
Hedge destruction 7, 117, 118, 121
Helmsdale 135, 137, 141, 142
Hen Domen, Powys 35
Hendre 91
Hen-house 41, 51, 83, 117, 118
Hemp 24, 41
Hereford 63
Heydon estate, Norfolk 76
'High' farming 8, 97, 102, 103, 105, 107, 112, 113
Hillforts 13
Holderness, Yorkshire 27, 32
Holkham, Norfolk 75, 79, 86, 88, 102, 103, 105, 148
Holkham sheep shearings 88
Holland 66
Hollowbank, Cumbria 73
Holme Moor, Devon 11, 47, 54
Honourable Society of Improvers in the Knowledge of Agriculture 95
Horses 23, 89, 93, 117, 118
Horse works 90, 92
Horticulture 115
Hound Tor, Devon 59
Hudson, John, farmer, Norfolk 107, 109, 112
Huntingdonshire 118, 120

Inbye 93
Infield-outfield 23, 25, 36–8, 42, 135
Irthlingborough, Northamptonshire 22
Iron Age 13–5, 18, 19, 24, 25
Irrigation 112
Isle of Axholme, Kent 35
Isolated farms 31, 40, 56, 57

Jacobite rebellion (1745) 95, 135

Kelp 97, 98
Kent 38, 63, 69
Kempstone, Norfolk 75
Kensworth, Essex 51
Kilkamkrill, Sutherland 143
Kilmarnock parish 96
Kilns, grain drying 16, 43, 51, 91, 136
Kinderley's cut, Cambridgeshire 126, 128

Index

Kintradwell, Sutherland 137–9, 140
Knapwell, Cambridgeshire 82
Knarr's Cross Farm, Thorney, Cambridgeshire 126
Knight family, landowners, Exmoor 121
Knights Templar 46

Lairg, Sutherland 143
Lake District, Cumbria 49, 51, 78
Lancashire 63, 72, 104
Landbeach, Cambridgeshire 56
Landlords 27, 30, 53, 58, 62, 66, 76, 97, 98, 104, 105, 107, 116, 135, 140
Lands End, Cornwall 12
Lath house 72
Laughton, Kent 38, 50
Lawes, John Bennett, agricultural chemist 101
Laxton, Nottinghamshire 8, 34–6
Lay subsidy 47
Leases 62, 95, 96, 107, 122, 127, 131, 137, 138, 140
Leigh Court, Worcestershire 47
Lekocetum, Staffordshire 19
Lentils 41
Leicestershire 63, 65, 78
Lichfield, Staffordshire 19
Lime burning 136
Liming 79, 95, 96, 122
Lincoln, Bishop of 30
Lincolnshire 54, 63, 77–9, 87
Linhay 72
Llanynys, Denbighshire 42
Loch, James, land agent 103, 133, 140, 144, 145
Lockinge, Berkshire 115
London 41
Longham, Norfolk 28–9, 79
Longhouse 17, 52, 59, 72, 92, 96, 100, 136
Loose boxes 116, 131, 138, 140
Loudon, J.C., writer 88
Lynchets 14, 16, 35, 54

Madder 66
Manors 25
Manorial courts 30, 32, 41, 51, 57, 59, 65, 68
Manor houses 50, 51
Maps 65, 75, 122
Market gardening 63
Marling, see Claying
Marsh 23
Marshall, William, writer 64
Marshland, Norfolk 47, 48
Marston, Lincolnshire 65
Matthew, Richard, farmer, Borlace 117
Meaux Abbey, Yorkshire 43
Mechanization 86, 97, 106, 118
Mechi, J., farmer 112
Melksham forest, Wiltshire 62
Mendips, Somerset 54, 64, 83
Messuage 55, 56
Milburn, Cumbria 31
Mileham, Norfolk 55
Ministry of Agriculture 117
Moats 7, 8, 36, 50–2, 146
'Model' farms 8, 87, 90, 92, 113
Monasteries and monastic estates 41, 51, 52, 59, 62, 68
Monmouthshire 63
Moors 12, 47, 78, 91, 92, 121, 141, 143, 145

Morris, William 46
Muchelney Abbey, Somerset 43
Mudford, Somerset 56
Mull, Argyll 95

Nadder valley, Wiltshire 67
Napoleonic Wars 78, 82, 92, 96, 97, 101, 108, 116, 121, 147
National Trust 46, 49
Nene river 124, 126
Neolithic farmers 9
Neolithic revolution 7
New Forest, Hampshire 40
Norfolk 15, 34, 37, 64, 87, 104
Norfolk Broads 80
Northamptonshire 63, 78, 83
North Elmham, Norfolk 25, 80
North Level, Cambridgeshire 124, 126
North Level Drainage Authority 124
North Level Main Drain 124
Northumberland 17, 24, 37, 63, 72, 87, 113, 121
Northumberland, Duke of 103
Nottinghamshire 13, 63
Nucleated settlement 23, 25, 27, 29, 30, 40, 41, 42, 51, 52, 61, 91, 100

Oats 24, 32, 41, 42, 54, 82, 93, 95, 127, 128, 135, 140
Olland 75
Open fields 7, 8, 25, 30, 31, 33, 35, 36, 38, 41, 42, 51, 52, 56, 58, 61, 62, 64, 66, 67, 75, 76, 82, 83, 91
Orchards 44, 62
Ordnance Survey 7, 8, 21, 52, 124, 131
Ouse river 68
Outbye 93
Outwell, Norfolk 43
Ovaltine Dairy Farm, Abbots Langley, Hertfordshire 117
Oxen 54, 92
Oxfordshire 63, 64, 77
Ox-team 13

Pannage 23, 29
Park Bow, Sussex 16
Parks 7, 50
Parliamentary enclosure 7, 38, 76–8, 81–2, 90
Partible inheritance 42
Pasture 113, 114, 127, 128, 132
Peak District, Derbyshire 54, 72
Peas 41
Peasant farming 16, 39, 52, 57, 59, 147
Peat fen 127
Peel houses 94
Pembrokeshire 14, 92
Pendock, Hereford and Worcester 50
Pennant, Clwyd 42
Pennines 17, 51, 78
Pershore Abbey, Hereford and Worcester 47
Pevensey, Sussex 23
Piddle river, Dorset 68
Pigs 9, 13, 23, 29, 39, 41, 46, 117
Pig style 51, 83, 92, 138
Pillow mounds 122
Pinfold 41
Plague 62
Planned farm 84–6, 90, 105, 138–9
Planned village 25, 30, 96
Plough 9, 13, 16, 23, 33, 34, 42, 90, 91, 93, 117

Population 9, 13, 16
Population decline 53, 57, 92, 107
Population increase 9, 13, 16, 17, 27, 32, 37, 40, 41, 47, 53, 58, 96, 101, 121, 135
Porlock, Somerset 41
Portland, Duke of 148
Potatoes 95, 98, 115, 135, 140
Poultry, see Hen-house
Prices 62, 63, 66, 78, 101
Puddletown, Dorset 68
Pulses 82
Pusey, Philip, writer/farmer 109

Quantocks, Somerset 54

Rabbits 121
Radnorshire 92
Railways 92, 101, 109, 113, 114
Ramsey Abbey, Cambridgeshire 43, 54
Raunds, Northamptonshire 18, 24, 25
Reaves 11–3
Reclamation 23, 43, 47, 51, 96, 122–3, 136–7, 144–5
Rennie, Sir James, engineer 128, 137
Rents 53, 82, 83, 101, 135, 140, 145
'Reversed-S curve' 34, 65
Rhiloist, Sutherland 144
Rick yard 92
Ridge and furrow 7, 32–5, 39, 41, 42, 44, 52, 54, 56
Riseholme, Lincolnshire 30
Rivet wheat 41
Roads 16, 79, 90, 98, 122, 136
Roman fields 8, 16, 38
Romans 13, 16–24, 32, 80
Romney, Kent 23, 43
Rosal, Sutherland 135–6
Rotations 13, 66, 82, 89, 107, 131, 140
Rothamstead, Hertfordshire 101, 114
Rousham, Oxfordshire 88
Royal Agricultural Society of England 101, 107
Royal Air Force aerial photographs 7, 21, 117
Roy's Military Survey of Scotland 96, 100, 135
Run-rig 93, 95, 98, 135, 137
Rye 17, 24, 32, 41, 95, 140

Saffron 62, 66
Sainfoin 66
St Leonards, Hampshire 46, 47
St Paul's, London 51
Saxon, see Anglo-Saxon
Scandinavian invasions 24–5, 27, 40
Scotland 16, 42, 93–100, 111, 112, 115, 133–145
Scotstarvit, Fife 16
Second World War 118
Sedgemoor, Somerset 80
Sefton, Lord 103
Segenhoe, Bedfordshire 30, 41
Sellar, Patrick, factor and farmer 137, 144
'Set-aside' 119
Sheep 9, 13, 14, 21, 23, 39, 41–4, 54, 58, 63, 64, 67, 85, 88, 91, 95, 97, 100, 107, 109, 114, 118, 121, 127, 128, 131, 141, 143, 144, 147, 148
Sheep fanks 143
Sheilings 50, 91, 143
Shelter sheds 83, 85, 89, 138, 140
Shilton, Oxfordshire 47

Index

Shinness, Sutherland 143
Shropshire 50, 63, 87
Shrunken villages 55–57
Shugborough, Staffordshire 88, 89
Silage 114–115, 117, 120
Simonsbath, Somerset 122
Skelbo, Sutherland 136, 140
Sledmere, Yorkshire 87
Smallholdings 141
Snowdownia, Clwyd 91
Soay sheep 14
Somerset 19, 37, 63, 119
Somerset levels 23, 43, 80
South Creake, Norfolk 84
South Eau, Cambridgeshire 125
Spalding 19, 43
Spaldwick, Cambridgeshire 30
Spelt 13, 14, 24
'Spinning gallery' 72
Spong Hill, Norfolk 25
Squatters 81, 91
Stables 8, 41, 51, 52, 70, 71, 85, 93, 131, 138
Stafford, Lord 87
Staffordshire 87, 103
'Statesman' farmers 72
Statute of Merton 50
Stanwick, Northampton 18
Steam power 86, 88, 90, 112, 128, 131, 145
Steam pumps 68
Stonea Grange, Cambridgeshire 19
Strathfleet 'mound', Sutherland 137
Strip fields 7, 8, 23, 50, 61, 75, 91
Studland, Dorset 16, 17
Suffolk 15, 34, 37, 63, 64
Sugar beet 114, 117
Sunk Island, Humberside 122
Surrey 38, 63
Sussex 16, 63
Sutherland 96, 133–145, 148
Sutherland, Duke of 122
Sutherland, Countess of 133
Sutherwaite, Cumbria 73
Sutton Bridge, Lincolnshire 128
Swaledale, Yorkshire 66
Sweetworthy, Somerset 22
Swinburne, Sir Edward, landowner 78
Sykes, Sir Christopher, landowner 87, 88, 89, 121

Taunton, Vale of, Somerset 64
Taylor, William, farmer 109
Telford, Thomas, engineer 128, 137

Tenants 20, 57, 58, 83, 91, 95, 96, 102–4, 116, 131, 138, 140
Teulon, S.S. architect 122, 129
Textile industry 57
Thames valley 14
Thiston and Rockingham forest, Hampshire 50
Thorndon Hall, Essex 84
Thorney, Cambridgeshire 43, 68, 122, 124–132
Threshing machine 86, 108, 109, 121, 131, 138, 140
Thurrock, Essex 18
Tiptree, Essex 112
Tittleshall, Norfolk 76
Torridge river, Devon 80
Torrish, Sutherland 143
Towns 16, 61
Townships 42
Trustees for fisheries, manufacturers and improvement 95
Turnips 66, 82, 86, 89, 95, 104, 140, 144, 147, 148
Turnip houses 86

Underdrainage, see Drainage
Upwell, Norfolk 43

Vancouver, Charles, writer 125, 126, 131
Vermuydem, Cornelius, drainage engineer 68
Vetch 13, 41
Villas 16–18, 23
Villein 41
Vills 8
Vineyards 44
Virgate 32
Voeckler, Augustus, agricultural chemist 102, 109

Wales 36, 41, 42, 50, 72, 77, 91–3
Walls 21
Wantage, Lord 115
Warwickshire 63, 78, 118
Wash, The Norfolk/Lincolnshire 19, 43, 47, 68
Wastes 40, 42, 43, 68, 76, 79, 92, 96, 113, 121, 145
Water meadows 67–8, 109, 148
Water wheels 92, 109
Watson, Lt. Col. David, land improver 95
Waveney valley, Suffolk/Norfolk 15

Weald, The, Kent 40, 51, 147
Weasenham, Norfolk 77
Wellingham, Norfolk 77
Welsh Folk Museum 92, 100
Wessex 12, 16
West Penwith, Cornwall 12
West Splean, Stirlingshire 16
West Stow, Suffolk 23–4, 29
West Whelpington, Northumberland 59
Wharfedale, Yorkshire 16, 35
Wharram Percy, Yorkshire 25, 26, 32, 58, 59, 60, 61
Wheat 13, 24, 32, 41, 82, 95, 113, 116, 127, 128, 140
Wheat Act 116
Wheldrake, Yorkshire 50
Wicca, Cornwall 12, 21
Widdowcombe-in-the-Moor, Devon 59
Williamson, Henry, writer 116
Wiltshire 67
Winchester, Bishop of 41, 58
Windmills 68, 88, 90
Wind pumps 125, 128
Wing, Tycho, land agent 128–9, 131
Wimpole, Cambridgeshire 33, 87
Wisbech 124, 126, 128
Witney, Oxfordshire 43
Woad 66
Wolvesnewton, Monmouthshire 92
Workshop Manor Farm, Nottinghamshire 88
Woodland 8, 9, 23, 29, 40, 43, 50, 66, 91, 113, 118
Wool 62, 63
Wool villages 56, 61
Worcester, Bishop of 53
Wyatt, Samuel, architect 84, 86, 88
Wychwood, Oxfordshire 50
Wynne Robert Watkins, landowner 92
Wynnstay, Clwyd 92

Yardland 32
Yeoman farmers 61, 62, 69, 89
Yeoman grazier 57
Yields 82–3
Yokelands 38
Yorkshire 18, 37, 51, 54, 63, 72, 77, 78, 82, 89, 104, 121
Young, Arthur, writer 75, 76, 81, 82, 103
Young, William, factor 137, 140, 144, 145

Zennor, Cornwall 12